The Bulb Hunter

TEXAS A&M
AGRILIFE
EXTENSION

Texas A&M AgriLife Research and Extension Service Series
Craig Nessler and Douglas L. Steele, General Editors

The Bulb Hunter

Chris Wiesinger & William C. Welch

Texas A&M University Press • College Station

General editors for this series are
Craig Nessler, director of
Texas A&M AgriLife Research,
and Douglas L. Steele, director of the
Texas A&M AgriLife Extension Service.

Manufactured in China
by Everbest Printing Co.,
through FCI Print Group
This paper meets the requirements
of ANSI/NISO Z39.48-1992 (Permanence of Paper).
Binding materials have been
chosen for durability.

LIBRARY OF CONGRESS CATALOGING-IN-PUBLICATION DATA

Wiesinger, Chris.
   The bulb hunter / Chris Wiesinger and William C. Welch. — 1st ed.
      p. cm. — (Texas A&M AgriLife Research and Extension Service series)
   Includes index.
   ISBN 978-1-60344-821-5 (flex : alk. paper) —
   ISBN 1-60344-821-7 (flex : alk. paper) —
   ISBN 978-1-62349-002-7 (e-book) —
   ISBN 1-62349-002-2 (e-book)
   1. Bulbs (Plants)—Heirloom varieties—Southern States.   2. Bulbs (Plants)—
Heirloom varieties—Propagation.   I. Welch, William C. (William Carlisle),
1939– II. Title.   III. Series: AgriLife Research and Extension Service series.
   SB425.W564   2013
   631.5'26—dc23
   2012043672

Contents

The Bulb Hunter

BY CHRIS WIESINGER

I did not understand this at the time, but there is a difference between an adventure and a quest. An adventure, according to J. R. R. Tolkien, is a "there and back again story." One leaves home for an adventure and, while gone, has exciting and daring experiences but then returns home to life as it was before. A quest is what happens when one leaves home, with the extreme possibility that there is no chance for return, and if a return is made, things have so changed that life will never be the same again. If I had realized this was going to be a quest, I might not have even begun.

Introduction: The Cabin

The piece of land looks and feels like a winter wonderland at times, with a gray sky, geese flying overhead, and ice everywhere leading up to the little seven-acre lake. Other days it could be a tropical rain forest in monsoon season, very wet, but hot and sticky at the same time. Your feet will begin to sink into the sandy loam soil as you are attacked by mosquitoes. Such was the morning I awoke on my cot in the little red cabin by the lake, just into my new business or, rather, my new life. The temperature had already reached a balmy eighty-something and was bound to rise into the hundreds within hours. I lay there on a cot in the middle of the main room, scratching my bug bites from the night before, reminding myself to watch a few that seemed unusually large in case they turned black or blue or some other color telling me it was the bite of a poisonous spider. Still lying there, I thought to myself that if Kelly would fix the plumbing and shower, this cabin wouldn't be half bad. Already I had taken his old shag carpet outside and burned it and moved the chair with mice living in it back to his sweet potato barn, which was just down the road. He doesn't come down to the cabin much anymore, unless to play poker with his friends or take his kids fishing, so he lets me stay here. Who is Kelly?

Kelly V. Hamrick is the sweet potato farmer who owns the cabin. He was the first person to lease me a small piece of land to farm. Later, when he found out I was sleeping in a small rented office in the nearby town, he offered this cabin. His grandfather had built it in the 1930s or 1940s, and at the time they used mules pulling implements to dig the dirt out for the lake next to the cabin. Both the cabin and lake were in a small valley in the heart of East Texas, a gently rolling land filled with pine trees and large old red oaks. This is a stark contrast to the flat prairies near Dallas, just two hours to the west.

Everyone assumes my little red cabin and the bulb farm are on the same piece of land, but the farm is just down the road, because there are too many cows by the cabin, and too many hogs come to the lake, rooting up everything in their path. Hogs (depending on the number) could easily dig up and destroy an acre of bulbs in one evening. The farm is a five-minute drive from the cabin, but it's not a bad work commute at all. That farm is home to all of the flower bulbs that I have collected from old homesites, abandoned lots, generous friends, historic places, roadsides, and many other places. Many flowers that have long been forgotten can be found in such locations. If you don't know much about bulbs, you are probably most familiar with common varieties like tulips, daffodils, and hyacinths, but there are many more. Collecting, farming, and reselling these persistent horticulture treasures is precisely how I make my living. As you can imagine, my story is just as much about the flowers as it is about the adventures I have had looking for them.

After working on the bulb farm, I would often come back to the cabin exhausted, covered in mud, and pull out the cot for my night's sleep. On my radio I hear the traffic reports in the mornings from nearby cities. Usually there is an accident on one of the major highways to Dallas or Houston. I smile at times, because on my little farm-to-market road, there are no bottlenecks or fender benders to slow me down as I twist and turn these country roads. Tyler, about forty-five minutes south of my little town of Golden, Texas, is growing, but the roads next to Golden are quiet. Not a bad five-minute drive at all, except that I might be a little tired from a restless night on a cot in the cabin.

Over my first few weeks of living in the little red cabin, the high temperatures began to drop from the hundreds to the eighties, and then within hours one day the high temperature was fifty

The little red cabin. CLW

degrees. The highs hovered in the sixties for most of the winter, with lows dropping into the teens (the low would usually be in the upper thirties or low forties). The first night the temperature dropped, I looked for the thermostat. Then I looked for some kind of wall unit for heat. I saw a gas hookup, but the propane tank outside was empty and there was no gas heater to hook up. I resorted to running around in the surrounding woods looking for fuel for a fire in the fireplace. Once roaring, it heated the cabin, but roaring fires eat wood fast, and around 2:00 a.m. every morning, the chimney would stop channeling smoke and begin sucking the heat out of the cabin. It went from comfortable to cold in about fifteen minutes.

I had always thought of myself as a minimalist and took only two sets of clothes with me when I started college (I had uniforms for daily life there), and this cabin fit into that theme. A long wooden table stretched almost the entire width of the main room, with a couple of wooden benches and a few wooden chairs. They were good for sitting down and eating but not for relaxing. To relax meant pulling out the cot that lay propped in the corner, opening it up, and lying down. In a little add-on room, Kelly had managed to install a bathroom and bunk bed frame that I ran a bar across and used to hang clothes. I didn't sleep on the top bed because it was too close to the ceiling and I feared the numerous spiders that would crawl in through the cracks in the evening. The water supply for the bathtub was a ten-gallon tank in the kitchen area, and that equated to about three inches of warm water.

When the sun comes up, it shines over the lake into the cabin, and if you walk out the back door onto the little deck in the mornings, you can generally warm up pretty fast. By that time I'm warm, the coffee is ready, and not soon afterward I begin my day. The blankets used on the cot go into a big Tupperware container with a little bar of soap at the bottom to try to keep them fresh because there is never time for the trip to the laundromat. The rest of my clothes I wash while staying with friends as I travel around looking for bulbs.

In such an environment, I often have the time to be dangerously introspective and ask myself questions such as, "How was it again that I ended up here?" I trace it back to my senior year as a horticulture major at Texas A&M University when I wrote down a simple idea to start a business. This idea was actually for a senior class project in which we wrote a mock business plan. After some thought, I settled upon a plan to offer perennial, historic, heirloom, warm-climate flower bulbs that persisted for gardeners in warm climates. This answered a long-lasting concern I had because of my inability to grow repeat-blooming tulips, and this seemed the perfect industry niche for what was surely the droves of disgruntled bulb growers. This niche in horticulture needed to be filled and led me to looking for flower bulbs and living in this cabin.

I would later develop an understanding of the connection between my own discomfort caused by the rapidly changing temperatures and the discomfort and death this weather could cause plants. This connection between my life and flowers would continue to increase; therefore, this story is best told by exploring the parallels between my life in the cabin and the life of the bulbs.

I'll begin with the business. As I was finishing the plan for my class project, I realized I could not be a broker for the heirloom bulbs: there were no growers in the warm climates around the world for the specific bulbs I needed. Those growers that did exist offered some bulbs that would work for us, but not all. As I struggled through this portion of my plan, I found myself gazing out the window of my truck down many old country roads. Looking out into pastures, I could see places where homes once stood, with their foundations, sidewalks, and garden beds now marked solely by daffodils. Upon closer inspection, I saw there were also heirloom gladiolus, hyacinths, rain lilies, crinum lilies, Easter lilies, and more. These were the bulbs I needed for my plan.

These old-fashioned varieties that persisted came from different areas. Many of those places were in the path of expansion projects for housing developments or new roads. Some were in old city lots, and others, in pastures. I started to reach the scary conclusion that brokering was out of the question, and there were no other sources. If my plan was going to happen, I would have to become a plant collector from these places, and then even more drastic—I was going to have to become a farmer.

My business was turning into more of a life-altering choice, and *farmer* was a scary word to me. I recalled the days of my

The Tauzin-Well House, the oldest structure in Natchitoches, Louisiana, is not an abandoned, run-down structure, but it does demonstrate the quality and quantity of heirlooms I had begun to see springing up in old yards. It is now surrounded with red spider lilies. CLW

The run-down boathouse that accompanied the cabin and lake. Dave Shafer Photography

youth working for an onion packaging operation in the Central Valley of California—it was hot, dry, and dusty. It also reminded me of the grim outlook about farming I received in my agriculture economics classes—farmers now pray for both rain and government-subsidized buyouts.

My business plan put the words *bulb* and *farmer* together, and everybody in the bulb industry responded, "Why?" Trade leaders told me about the falling market. They mentioned that the only reason bulb distributors were still around is that big-box store expansion was pulling the demand, but more and more bulbs were simply being thrown away every year. Fewer and fewer people were buying bulbs. When I asked why, they would respond that we live in an instant society; nobody wants to wait for food, service, or garden color; and they want their gardens to look pretty now. It appeared the misery of farming loved the company of flower bulbs in a slump market.

When most people were exiting the farming and bulb world,

I entered it. In the fall of 2004 I officially started what would be called the Southern Bulb Company with a twenty-five-thousand-dollar investment from my first limited partner. I had a disadvantage in that I knew nothing about business, nothing about farming, and very little about bulbs, even with my horticulture education. All I knew was what type of soil I needed—a sandy soil for easy harvesting of the bulbs, just like a sweet potato farmer might. Most of the sweet potatoes in the state were grown in the sandy loam soils of northeast Texas, so I drove to Wood County and met with the local extension agent, who gave me a list of sweet potato farmers. He said that if I went to Domino Hall (a restaurant with tables for domino playing) in the town of Golden at lunch, I could probably find them all playing bones or 42—the traditional games. I decided to call first (rather than make a proclamation to the whole restaurant of my intent to grow flowers in Wood County) and picked a random name from the list of sweet potato farmers. Kelly Hamrick looked good, so I called Kelly, who was not at the restaurant but out at his barn boxing up sweet potatoes.

I found him among the crew of five Hispanics, working the packing line. Kelly was making sure he was dumping enough large bins of sweet potatoes into the washing basin to keep the potatoes moving down a conveyor belt. Once the potatoes were washed and sorted, the crews would throw them in bushel boxes for shipping. The water from the washing basin drained out the back of his barn, through a muddy pen of cattle, and into his farm pond. Beyond that was an expanse of green acreage with slightly rolling terrain that no doubt belonged to him.

During the basic introductions, I could hardly hear him over the machinery but did catch his words when he finally asked what I needed from him. The problem was that I didn't know what I needed. This whole venture would have been easier if I knew what I needed. I blurted, "Just a few acres."

He thought as he worked, spit tobacco, and looked at me, "Just a few acres?"

"Yes, sir, not much."

"I've got a spot, across from my mom's. It's kind of set back and where the hogs wouldn't get to them. Do you know where the road curves just past that . . ."

He continued, but I zoned out. I didn't know where his mom's house was, or where the curve was, or much more about the exact location. All I knew was that he allowed me to use a piece of his land. I thanked him profusely. A week later, I took my hoe and bulbs, called my brother, and we began to plant our first bulbs on a ten-acre plot of sandy loam soil.

I also looked for and rented a small office in town, but two months later Kelly offered me his little red cabin on the lake. It was a major victory. I now had a place to stay right by the farm, and it was just what I needed: shelter, water, and food (bass from the lake). This sweet potato farmer had allowed me a place for my business and bulbs to take roots, and now I was putting down my own roots.

CHAPTER 1

A Little Red Tulip

My story began with a red tulip and is how I begin my talks to garden clubs and organizations:

"When I was young, my family moved to California. We had gardens everywhere we lived, from my birthplace in Lafayette, Louisiana, to our old hometown of Houston, Texas, to the very hot town of Bakersfield, California. Most weekends in Bakersfield were spent maintaining the yard and visiting the garden centers," I continued. "One fall weekend, when I was about eleven, I purchased an object that looked something like a rock but was displayed in a cardboard box that had beautiful pictures of flowers all over the outside. I planted my little rock in the ground and forgot about it. Four months later, something special happened.

"It was one of those few transition days in the two weeks between winter and summer that we in the warm climates call spring. (This statement always seems to tickle the audience.) A rain shower had cleaned the air so that we could actually see the Sierra Nevadas and the Coastal Range looming in the distance, reminding us that we lived in a valley. Rays of sun pierced through the clouds and the rain had stopped, so my parents opened both front and back doors to allow a pleasant draft to sweep through the house. It was in the middle of this draft that I lay napping on the floor, occasionally staring out the front door. My eyes caught a glimmer of a spectacular color out in our front garden, and I went to inspect it. Something magical had occurred;

A single red tulip. CLW

my living rock had turned into the most striking red tulip. To me, it was a miracle, and gardening had come alive!"

I then began describing the characteristics of flower bulbs. "According to August De Hertogh in *The Physiology of Flower Bulbs*, all bulbs belong to a category of plants called geophytes— plants that survive by using their underground storage structures. A true bulb is similar to an onion, but many 'bulblike' organisms such as gladiolus (which are corms) are almost always mislabeled as bulbs. Lilies, tulips, daffodils, and dahlias all have different structures, some with scales, tuberous roots, and so on, and have different methods of propagation."

I began to lose the audience as I saw their eyes wander, and I heard a few people cough. I went into panic mode. What was my purpose here this evening—to educate or entertain? The material on the botany of a bulb was good information, but a dry presentation would not lead to a successful evening talk. I should have known this when I first walked up and my outfit was critiqued. Someone told me, "This is a garden crowd, so you can lose the tie."

I quickly realized that my life wasn't a dry presentation; I had been on adventures, been uprooted, and had moved to a tiny red cabin to farm bulbs, so why should my presentation be dry? Quickly, I went back to more personal stories, intermingled with botanical tidbits.

Continuing the red tulip story, I explained my hopeful anticipation for the next year's bloom. I eagerly watched the foliage come up on my bulb, but there was no bloom. "That's all right," I said, viewing myself as a patient child. "I looked forward to next year's bloom. Next year came, and there was not even any foliage. Once I had dug down, I found the rotted remains of my bulb. It had died. I was horticulturally scarred! (Generally this receives a few laughs.) Some serious questions arose about why this had happened, whether or not I had done anything wrong, and if anything could be done in the future. My quest for knowledge led me to realize that some bulbs were not meant to come back every year, that they were annuals to be enjoyed once and that was it. I found an appreciation for such bulbs. However, I thought to myself, 'Surely there must be bulbs that do come back every year, bulbs that don't need to be dug every year and that will grow, multiply, and bloom in warmer climates.'"

That is now the introductory story I tell to gardening groups. The red of that first tulip is still in my mind, and I connect it with falling in love with gardening. I also connect it with other things, such as a peaceful afternoon, my mom, and of course the beginnings of my business. Nine years after my tulip encounter I entered my first horticulture class at a junior college in Bakersfield. One transfer and four years later, I graduated with a horticulture degree from Texas A&M University in College Station. I was about to be introduced to a world that I couldn't dream existed and thrown into a flower bulb industry to face many personal trials.

To begin either a quest or an adventure, we need someone to push us out the door. That person for me was Bill Welch. After a couple of weeks thinking about and writing that early business plan project, I knocked on Bill's office door. He was a professor who was considered an authority on heirloom plants, including bulbs. I did not expect what happened after he had listened carefully to my plan. He responded with pure and excited encouragement. I had never felt such validation before. I asked for another visit, and he welcomed it.

While a horticulture student at Texas A&M, I also participated in the military tradition of the school as a member of the Corps of Cadets, with about two thousand full-time cadets who lived in dorms on campus. Cadets wear the uniforms that the army wore in WWII, have a complex student leadership structure, and live with the disciplined life you would expect of such an organization—waking up for runs in the morning, marching to meals, living in close quarters in the dorms, and so on. I was in this organization but found time to work in the greenhouses during the day. My nickname among the guys was an obvious choice: "Flower."

In school, I often stood in the back of classrooms so I wouldn't fall asleep. It was poor manners for anybody, much less a cadet in uniform, to fall asleep in the middle of a lecture. In one such class the horticulture professor spoke about possible ideas for the main project of the senior-level course, and I was listening because the task was essential for graduation. "The business plan could be to

2004 5 15

grow trees, perennials, open a tissue-culture lab, anything nursery/garden related, or it could be about bulbs." Flower bulbs had always been at the root of my love of horticulture.

As I walked from class to the dorms, new thoughts swirled in my head. While waiting for an afternoon Corps meeting to start, I mentioned my business plan to sell flower bulbs to some of the guys with me. Slightly surprised, one looked at me with a sheepish smile and asked, "What?" A few others asked some complimentary questions, but then it was time to begin our meeting. We finished the training run and broke for dinner. That evening in the dorms, I kept my confession to a few of my closer senior friends. They asked if I was serious. "Yeah, I guess so. I just need to look at it, but I think I might really do it."

To look at the numbers was to become depressed. I used limited information to run the numbers for the business plan. All I

Anybody in the military will notice slight technical problems in our marching, but this was a great group of men! "Flower" is in front. CLW

really had was the National Gardening Survey and some leftover statistical books I had taken home with me from Washington, D.C., after a brief summer tenure with the US Department of Agriculture. Flower bulbs since the 1950s had done extremely well, but in 1995 we began to see a sharp decline in the industry. In fact, every year since 1995, sales have decreased by 3 percent annually and have to this date not recovered. How this has affected Dutch and US farmers and US bulb businesses is harder to say. Many bulb businesses are still privately owned, and nobody on the outside looking in really knows there is a problem until the businesses go bankrupt.

In my sleep I began to turn the numbers over, trying to comprehend the unknown and come to conclusions. I had several more visits with Bill. My thoughts were moving out of the theoretical to the practical, and he began offering some advice. He gave me names of farmers, historic associations, contacts, horticulturists, etc. that I had to visit, see, and know before I could ever hope to embark on such a mission. As my last visit with Bill before Christmas break concluded, I did not realize that he was doing more than ushering me out of his office; he had just nudged me into my first bulb adventure.

That spring break I took a road trip to begin meeting people and asking them questions, and I also began to call world sources for bulbs and tell them what I wanted to do. The responses of larger sources were grim: "I don't think there is a market for those bulbs!" or "There is no way you can farm those bulbs and make a living." I began to doubt the possibility of such a business, but there were glimmers of hope here and there on that trip. As I drove around Texas that senior year in the spring of 2004, I sometimes found myself on old country roads. It was on this trip that I first noticed the sparks of color in the old pastures, where homes used to stand but were long gone, and lines of daffodils bloomed to mark old foundations, garden beds, and driveways. These were the bulbs I fell in love with and that I needed; these old bulbs were the future of what would become the Southern Bulb Company.

I was hopeful about my idea and upon graduation hung up my uniform and traded it for the muddy jeans that would be indica-

The daffodils in this pasture were no doubt spread about by tractors tilling what is now farmland. CLW

tive of my life as a bulb collector. My new nickname among those in the industry was becoming "Bulb Boy," perhaps a step up from "Flower."

Collecting bulbs and stories, I knocked on doors, made phone calls, and tromped around pastures, woods, and roadsides. There was still something missing: a tulip. Maybe, just maybe, as I transitioned into a life of searching for bulbs and living in a little red cabin by a lake, I would find the answer to the red tulip of my youth. Could I actually find a tulip that would come back year after year? Would I find it along with other heirloom bulbs in an abandoned site? If I did find it, would other people be able to grow it?

It was very nice to have my older brother, John, with me at the beginning of my business. He is someone I have always looked up to, and he remains my best friend. He went on many of the early adventures of bulb hunting with me, such as one winter morning in 2005.

We left in John's midnight blue Ford truck, its headlights piercing through the early-morning darkness. It was a cold Texas February morning, a morning that was ushering in a cold front. Wind and a high overcast sky yielded a little rain, which tried to hinder the vehicle's progress as it slid through sleepy country towns. Inside the truck, climate controlled, the radio on faintly in the background, all was well. In my famous manner and to John's chagrin, I quickly fell asleep.

There was the Bulb Boy, twisted and contorted in such a manner as to make the thought of someone being able to sleep, and sleep soundly no less, impossible. John was forced to rehearse the directions in his mind again as we approached the town where our potential find was awaiting. In the back of the truck were the tools modified for the trade of adventure and bulb hunting—special shovels, picks, crates, and gloves.

As the first rays of morning light began to illuminate the backs of the clouds, I finally began to stir, subconsciously knowing that we were closing in on the town where the prize lay close at hand. A few turns led us past very inviting diners, but we knew that what awaited us was a cold, drizzly morning full of physical labor. As we drove over crushed gravel, we passed chain-link fences and abandoned cars on cinder blocks that had replaced white picket fences and manicured gardens. Finally, the truck stopped, and the only words that came to me were, "There they are!"

We both sat in the truck longer than we should, taking in as much heat as we could before going outside. Finally, John watched as I opened the door and walked to the back of the truck for my tools. We stepped out of the truck into deafening silence. There were no birds chirping . . . no sound of life. We could hear only the wind and the light frost crunching beneath our feet. We both walked cautiously toward the "abandoned" home, knowing that many of these homes, dilapidated as they are, can still be refuge to folks eking out lives for themselves. No one was there, and the address confirmed that this was the house where we received permission to dig.

John followed my footsteps in the frost to the side of the

house where the contrast in colors on the ground became much clearer—vibrant emerald greens and sapphire blues. The frost on the foliage and petals began to glisten in the emerging morning light. The scene is the same all across the South in the two to three winter months, and to be clear, the "South" generally includes the southern US states and Texas. At first glance, it looked like just an abandoned home, but in it and around it were life, art, and beauty. The blooms, fragrance, and generations of gardeners represented by those flowers bring warmth among the frost. This warmth is what infuses life into the crowds I speak to, putting smiles on strangers' faces by continually finding blooms amid bleakness. These flowers are eternal optimists. John, a commercial pilot for Delta, began to see the art and realized the magnitude of the life project that this was about to become.

My transition from college life, my move to the cabin in East Texas, and some of these very first bulb hunts all occurred in the fall/winter of 2004. In the late spring of 2005, a year into the business, my brother continued to believe in me. Considering he was a commercial pilot, I don't know why this should have instilled any confidence in my plan. He offered to help me plant the field that year to assist with my increasing workload. A historic garden group called needing a speaker, and my talk, one of my first speaking engagements, fell on the day we had set aside for planting. John sent me on my way and said he would take care of the planting. What a relief!

I needed this help; I was exhausted. Nobody wanted to repeat the mistake from the year before—I had planted the newly rescued bulbs too shallow. The 104-degree-, full-sun-, no-rain July killed many of them. The bulbs were not as foolproof as I had thought. Most bulbs, like daffodils, need to be planted about two to three times the height of the bulb below the soil surface. Some bulbs, like crinum and amaryllis (*Hippeastrum*), are different, but most fall-planted/spring-blooming bulbs can be planted according to this rule. Unfortunately, the rule is sometimes misapplied, and we have had several customers dig holes two feet deep for their larger crinum bulbs.

As I arrived at my talk, John called:

We can drive John's truck right out into a pasture next to the clumps of bulbs. Featured here is the 'Campernelle' (Narcissus x odorus). CLW

"Hey, Chris. It's John."

"Hey," I responded from the little historic center where I was about to speak.

"Got the bulbs planted."

"Great! Thank you, thank you."

"No problem. We planted them a foot and a half deep."

"Okay," I responded, my voice changing ever so slightly as I drew out this word just a little longer.

It was my brother I was speaking to, and he of course caught my tone. I had betrayed my anguish, my despair, all in an instant. We would never be able to completely harvest these bulbs at that depth unless we spent ten minutes on each bulb excavating them with a crew. I quickly tried to recover, expressing my thanks so he would not feel bad. It was a problem, but many of the bulbs did survive. I now can tell audiences that the bulbs are pilot-proof!

We both kept our energy and enthusiasm high. John was extremely adamant about some things, and one of those was having a tulip. My enthusiasm to find a red tulip started me on this adventure, but after more disappointments I felt compelled to explain to my brother that tulips don't do well in Texas. I had already had my bad experience as a child. Why embark on another disappointment? John's excitement persevered and went with us on a trip to visit a well-known horticulturist in the small East Texas town of Center.

Most small Texas towns have a Mexican food restaurant. Some are better than others, some serve margaritas, and others are in dry counties or just could not afford the liquor license.

"Margaritas or beers?" someone asked our waiter.

"Nooo. I'm sorry, sir, but we don't have any."

"Dry?"

"Yes," he said with a smile and expectant pause.

"I'll have a Coke, please."

Orders went around the table to my brother and to Greg Grant, our host that day, who was considered one of Texas' best horticulturists. After we toured his early twentieth-century home (which had once belonged to his grandmother) and bulb farm, he had brought us to this restaurant. The terrain in Center was much like that close to my little red cabin, except many more pine trees filled the woods, and its economy was larger, with its vibrant logging, oil, and gas industries.

It had been a long day, and my brother had absorbed some of the information from the tour, but he leaned forward and stated what he felt was the most important point of the trip: "What we really need is a tulip that does well in Texas."

John started with Atlantic Southeast Airlines and now flies for Delta. The passengers might be surprised if they knew their pilot helped farm bulbs in his free time. CLW

A slight pause followed . . . some eating and looking at the plate.

"Well, there is one tulip that might work. Some people say it can be found in a little town south of Dallas. Jimmy Turner at the arboretum knows more about it." Greg was referring to the Dallas Arboretum, and Jimmy Turner definitely knows about tulips. If you ever want to see one of the best tulip displays, if not *the* best tulip displays in the South, visit the Dallas Arboretum for its "Spring Blooms Festival."

I visited Jimmy shortly after the visit with Greg. He spoke about a particular tulip he had known in the small town where he grew up, and it was in one of the five cemeteries that existed there. Jimmy rarely returned to his hometown because he was now the ever-busy director of plant research for the arboretum. He offered to give me some of these tulips. (Tulips will be covered in more depth later, especially in the section "A Primer.")

The tulip had stopped blooming by the time I saw Jimmy again, but he had some bulbs for me in a little bag. He also gave me some leads on where I might find more. Then in mid-March, John called. I was already wrapped up in other spring activities and had almost forgotten about the possibility of tulips. At first there was some general chitchat about things and then the question:

"Are you going to that town to look for the tulip?"

"I would, but I don't think it's time right now, and I'm swamped with things to do."

"It's mid-March, isn't it?"

"Yes."

"Then it's time to go look for the tulip. Chris, do what you think you need to, but it's the tulip."

I hung up the phone, took a breath, and sat in silence. With a step out the cabin door, I left on my adventure. This was to be one of many adventures that together have led to a life-changing quest that eventually earned me the nickname "The Bulb Hunter."

CHAPTER 2

People Are Like Crinums

In the late spring and early summer of 2005, my business was to the point that either it needed more funds to continue growing or it needed to be shut down. Searching for the tulip continued but so would hunts for other hard-to-find bulbs. My hunt for bulbs and my hunt for money to keep me afloat went hand in hand. I shared with John my adventures for both, and he continued to ride along with me on bulb hunts and was always inquisitive about the bulbs:

"How much can you get for that?" he would ask.

"That? Oh, that's a crinum. Some people really don't care for them." I would pause and look over at him, "But some really

Close-up of a bloom of a milk and wine crinum (Crinum × herbertii). CLW

do. They can bring in twenty dollars, thirty dollars, sometimes forty dollars a piece." Greg Grant had given us an introduction to crinums, but a dollar value was something anybody without a horticulture education could understand. By the end of this chapter, you the reader can determine whether or not the value of a crinum lies in its demand or in its supply.

"Crinum" is both the common name and botanical name (the genus *Crinum*) of the bulb. An elephant of a bulb, it is grounded by thick, wormy roots but above the ground creates lush foun- tains of foliage that can become home to other little creatures. In some varieties, the foliage naturally tapers to a point and curves around at the tips. This sensitive area can brown in the heat of the summer, making them unattractive to some people. Crinums propagate quickly and will firmly establish themselves in a solid clump after a few years. "Solid clump" is putting it mildly.

Homer wrote in the *Odyssey* of a flower that even the gods

On top of my old truck, the roots of the crinums are displayed to show their size in comparison with other flowers. CLW

My pile of broken shovels from trying to dig crinums. CLW

could not uproot. He must have been talking about a crinum. If you try to remove crinums after years of growth, you'll probably end up with a pile of broken shovels. I have just such a pile on the farm: cheap shovels, expensive shovels, sharpshooters, regular shovels. The best shovel rose to the top: we affectionately called it the "Structuron." It was almost indestructible, but it still broke when John and I pulled on it at once.

When I was invited to speak to a Master Gardener group near the Texas coast, I was also invited to visit a personal garden and dig some crinums. I accepted the invitation. Before the talk, I met a few members at the last crossroads before the small road we turned on led south to the little coastal town of La Salle and then into the Gulf of Mexico. I arrived at our spot and met the group. Next to my truck we had a discussion about how hard it can be to dig crinum bulbs.

"Didn't I tell you I had the answer to that problem?" my host, Janie, remarked.

"Yes, ma'am, you did."

"Well, I keep telling my husband that we need to patent it. Do you know anything about patenting?"

I knew a little, not much, so I just responded, "Not very

much!" I was smiling a little bigger than usual, trying to mask that I was tired from the six-hour drive I had just made from Wichita Falls in northwest Texas to Houston and then to their town, but also trying to signify that I actually did look forward to seeing this mystery tool. We loaded into our cars and embarked for La Salle.

At each turn I saw more water, and soon water was on both sides of the road, yet we continued south with no scenery but scrub brush and native coastal grasses as far as the eye could see. It was so different from the towering pine trees and sandy loam soil that marked my area of the state and most of the Southeast. We kept driving and driving, closer and closer to the ocean, with canals now to the left and to the right. I thought La Salle might actually be located in the ocean, but eventually we came to an area where the land opened up, and I spotted a row of houses in the distance. We came to a sign: "La Salle: Population 17."

The town is named after René-Robert Cavalier, Sieur de La Salle, the French explorer who set out to establish a French colony on the Gulf of Mexico. His ship's crew met a bad fate, but as a result of his exploration, the French flag is one of the six flags that have flown over Texas. I assure you the Texas coast is much friendlier now; however, the coastal plains are still a rough environment for plants. Crinums, of course, love it there.

We arrived at our destination, and I knew from this garden spot that the gardener Janie was about to introduce me to was a character. I often found myself playing Sherlock Holmes, trying to judge the character of people based on their gardens. In this case, the biggest clue was the large number of crinums in the garden. This would be a good time to mention why some people do not like crinums.

Crinums are natural, free-flowing forms of beauty, but some people use the less friendly term "messy." Most gardeners these days want the flowers only, not the year-round foliage that goes with them. For example, violas are mostly flower with very little foliage visible. When the flower has stopped blooming over the winter and early spring, the plants are simply pulled up and thrown away. They are extremely popular and are seemingly in every garden in the South. Crinums have large, impressive flowers but also have a lot of foliage. The two- to five-foot-long, bold foliage often buckles in the middle and flops to the sides of the bush, giving it an unkempt appearance. A gardener who incorporates crinums into the garden, either successfully or not, definitely breaks the modern mold and is what I call "a character."

This person's home was built on stilts like all of the other homes there. Her garden was full of plants, plants, and more plants—growing everywhere and growing big. They were all "tropicals," and tropical plants can be aggressive. In warmer climates, many things grow faster and larger than in more temperate climates, and aggression is a key goal to survival. The homes on the other side of the street didn't have lush backyards but were landscaped with mowed lawns that sloped to boat docks on the tributary that led to the ocean. This clean appearance and low grass helped the residents spot alligators.

Our garden host came down the stairs and began shouting greetings along the way. She was well tanned from more than fifty years spent in the sun and had a big smile. The smile paled in comparison to the personality. We soon moved on to the plants she wanted out.

"I hate 'em!" she exclaimed.

"We love them!" the other ladies chimed in.

"Well, I have to dig them," I thought to myself, but not a bad trade-off for a truckload of bulbs. Now, I wondered, what is this tool? Will it really end all digging problems with crinums? Janie pulled out of her car what appeared to be a shovel.

"Looks like a shovel," I coolly stated with an attitude and look used only with a friendly group of ladies who had so quickly allowed me into their inner circle.

"Well," she started, and I knew she planned to dish out an equally sassy volley. "It *is* a shovel." She did not add a descriptive adjective of me at the end of her statement, so I assumed our relationship was still friendly. "But my husband was tired of breaking shovels, so it is the head of a shovel welded onto a steel bar."

She continued her description and pulled out the second part of the tool. It was a three-foot hollow, tubular implement with handles on the side—something usually used to pound stakes into the ground. One end of the tube is open, and the other side is sealed shut. The open end of the tube slides over the stake (in this case the other end of the shovel), and the handles are used to slam it down toward the ground until the end of the stake hits the solid end of the tube. It is kind of like a hammer for long skinny objects and in my experience is used often to drive metal fence posts into the ground.

She began to drive the shovel head under a large patch of crinums. I could get a shovel under a patch of crinums, but it usually took awhile. I saw the first signs of a time-saving device, and I became interested. I looked at her pole and shovel now under the clump. "There is no way she will be able to pull up this clump," I thought to myself, recognizing the strength of the root system. She turned to me and said, "Here is the secret weapon." She turned and yelled toward the house, "Come on, ladies!" as five other women came over and all found a place to pull on the long pole, resulting in amazing amount of leverage. Up popped the clump, and I was duly impressed!

After a while, everyone went upstairs for cool drinks on the screened-in porch, but I continued to dig. It was hot, but luckily it was still "winter" and the mosquitoes weren't bad. Once I finished, my hostesses came down, and we all agreed to see each other that night for my talk at their Master Gardener meeting.

Talking about crinums to the Master Gardeners. WCW

I hopped in the truck and drove back to my hotel for a quick shower and change of clothes. With me was a load of crinum bulbs with some other bulbs I had brought down to sell to the group—selections that could withstand the coastal weather and bloom every year with no chilling. I was going to show them pictures and tell the gardeners all about them. My talk that evening included a section on crinums for sure.

Crinums grow in ditches, swamps, and open fields. When Hurricane Katrina struck New Orleans, one particular house was inundated with twenty feet of salt water for two weeks. The only plants that survived were the "Peggy Martin" rose and a crinum. Crinums are survivors, and they have a wonderful variety of blossoms that bloom during the worst times, such as the hottest part of our summers.

Some of their lily-shaped blooms, which grow on three- to four-foot stalks, are pure white, others are dark burgundy, and some are yellow (although extremely rare) with all shades in between. And they have just as many variations in fragrance. Some, such as C. 'Mrs. James Hendry,' have a white flower with a light pink throat, a lovely fragrance, and clean foliage that answers the normal objections about their messy nature in the garden. Some are adapted to colder climates, and some tolerate partially shady areas, like C. 'Ellen Bosanquet.' It has a solid, dark pink, almost lavender bloom; makes jumbo bulbs; and reproduces quickly. The foliage can be a bit much for some, but with some simple companion plants around the base of the bulb, this isn't a problem.

Crinums most often seen in the South are C. *bulbispermum.* It is the first to bloom and the one found in ditches, hence the common name "ditch lily." Lightly striped pink and white blooms can be seen as early as April, but depending on the population or grouping of bulbs in any given area, this coloration can vary. An identifying characteristic is the wide, bluish-gray foliage that tapers almost to a fine point. It gives a striking appearance in the landscape, but the blooms remain small and somewhat closed so can be a disappointment to some gardeners.

Few crinums are native to the United States, and most originate

▸ *The pure white blooms of a* Crinum powellii *'Album.' CLW*

in West Africa, although others exist in Central and South America, Australia, Indonesia, on the Pacific Islands, and down into the Micronesian Islands. Crinums thrive in many different conditions and are extremely tough. They can be cut up and left on the side of the road with nothing more than ditch water to survive. Their tenacity, sometimes unattractive foliage, and beautiful redeeming qualities could also describe a person that meant a lot to me in my first three years of business. His name was Joe Bradbury, and he reminded me of a crinum.

Joe Bradbury worked for the CIA long before I knew him. He was a horticulturist working for a "fruit company" in Central America. This was, of course, a cover job for a much more important role he played for the United States. Don't get me wrong; Joe did know how to run a horticulture operation, but he was recruited out of Louisiana State University (LSU) to be a CIA operative. His stories about "kill or be killed" in Central America told over whiskey were most likely true, as verified by several credible sources.

In the 1960s, while working for the CIA, he was stabbed and left on the side of the road, drinking nothing but ditch water to survive. After a few days, Joe finally managed to roll out of the ditch and be noticed by a truck of Central American farmworkers. They took him to safety, and he made it home. I assumed the story was true because one night after the described conversation, he showed me his scars. After a few more drinks, I could barely keep my eyes open, but that is when Joe had just gotten started.

"Son, I've been following you since you were a sophomore at Texas A&M. I started to grow flower bulbs at Lone Star," he exclaimed as he glanced at the other person sitting at the table with us. He looked at me as if I were supposed to respond somehow. I shrugged and wrinkled my forehead in a feeble effort to show some signs of coherence and curiosity.

"Drugs and prostitution are the only two easier ways to make money than growing these bulbs. We were going to do it, but when Color Spot bought my nursery, they dropped the idea because it was a regional program. They didn't know what they were doing." Color Spot remains a large and healthy nursery to this day.

Joe started Lone Star Growers in the 1970s in San Antonio. His partner was a San Antonio oilman who noticed Joe working a landscape business on weekdays, weekends, morning and night, recruiting his kids to help him with the business and doing whatever it took to be successful. The partner decided Joe was the kind of person with whom he wanted to do business. Joe and his partner did just that by using only a small initial investment. The investor floated the company loans when necessary, and Joe was diligent about paying off the note.

Joe Bradbury started Lone Star Growers, a prominent nursery near San Antonio, Texas. Patsy Palme

Lone Star Growers became a large, growing operation, and when Joe was involved, it actively sought new plant materials from Mexico and throughout Texas. In the early 1980s he hired a Texas native plantsman icon, Lynn Lowrey, to search for new plants to introduce to the market. They sold to retail nurseries across Texas and the surrounding states. Joe understood the time it took to bring a new plant to market, from growing one plant to growing thousands, and knew the time it took to educate garden centers and the public to become accustomed to a plant. He began many regional programs matching specific plants to specific climates in areas with growing markets. A large customer base as well as loyal employees developed.

It was kind of hard to miss Joe. It was even harder not to hear him. I'm sure in those days his face was clean shaven, and he was probably skinnier. He probably didn't drink as much or eat steaks every night for dinner. He was a mental and physical warrior in the business world and had the charisma to pull it off. Not two months after he sold his business for $37 million in 1998, his wife died unexpectedly from a brain aneurism. Joe had a hard time recovering.

A mentor in college by the name of Buzz introduced me to Joe my sophomore year at Texas A&M. Knowing that Joe was a horticulture major at LSU and had developed a successful business, he got us together. Joe showed us around the old nursery and pointed out that right next to Lone Star Growers was another nursery that started about the same time he did. That nursery was barely limping along, with broken fences and overgrown fields. By showing me the difference between this struggling operation and Lone Star, he was telling me that I needed to listen to Joe. I would visit with Joe more and more often, and he stayed up to date with my plans. That was important, because Joe would later play a vital role in the Southern Bulb Company. A few more words about crinums are important, but then we will need to return to Joe to tell the story of the company and my life.

Tropical crinums include some of the largest and grandest, like *C. asiaticum*. Most of the "bulb" is aboveground, but the bulb is actually a trunk that looks like an enlarged, thick root, with smaller roots coming out the bottom of the bulb. The foliage can grow more than seven feet tall, essentially a small tree. The tips of the white petals on the blooms open up wide in large clusters on top of very tall, strong scapes. Scapes are the thick stems that come up from the bulb and support at their top the numerous individual flowers. The blooms fade, and seedpods form and eventually detach. Because the pods can float, the seeds can move across the water from island to island in the ocean. This bulb will do well all over the Gulf Coast of the United States, as well as Central and South America, California, Florida, most islands of the Caribbean, and many other tropical and semitropical places around the world.

The irrigation of modern home landscapes does not bother tropical crinum bulbs, nor does it bother the more temperate crinum bulbs. When most bulbs would rot for lack of good drainage, crinums do not mind wet feet. They can also be found looking green in landscapes that don't receive a lot of water. Other landscapes that might have problems with armadillos or squirrels seem to be just the spot for a crinum. I laugh at the thought of

Me with Crinum asiaticum, a very large crinum that can grow to seven feet tall. Bulbs in the South include more than just daffodils! CLW

those arrogant, bulb-digging squirrels attempting to uproot a crinum. For all of your friends who may not have a green thumb, give them a crinum and they'll think they are doing something right for once in their garden. For most of the common gardening ailments, crinums are the cure, but there are some drawbacks.

Like most bulbs, crinums die back in the winter, and the large amount of yellowing foliage can be considered an eyesore, but it is a survival tool. When the farm froze for an extended period of time one January, my pipes froze and cracked. The result was not a fun repair job, but most of the crinums, some of which were pushing the envelope of cold hardiness, recovered just fine from the extra blanket they received from their dead foliage. I was afraid my *C. asiaticum* would not make it, but in April most began to sprout. Some were lost, and a friend in Lafayette, Louisiana (which is even farther south), lost some as well that cold winter. My crinum collection continued to grow, but the eccentricity of the plants pales in comparison to the eclectic group of those who collect the bulbs.

Crinums seem to attract unique connoisseurs, such as the seventh-generation South Carolinian Jenks Farmer, who manages a crinum farm and collection in South Carolina. The Sons of the Confederacy tried to recruit Jenks to their southern heritage organization, but he continually told them over the phone, as he looked in the mirror at his yellow Mohawk, that they probably did not want him. I liked Jenks; he showed me around Charleston, South Carolina. I enjoyed viewing the private garden he was managing in Charleston, but I really enjoyed it when he took me to the best seafood joint in town.

Other icons in the bulb business include Marcelle Shepard and Steve Lowe, both in Texas. They sold me some of my first crinum bulbs to resell on our Web site. Marcelle lives near Beaumont, and Steve Lowe lives near San Antonio—completely different climates, but the majority of crinums don't seem to care. Both have large collections in their backyards, if "yard" is really the right term. Marcelle's yard backs up to one of the many bayou swamp areas of the Gulf Coast, and Steve's backyard consists of large bald cypress trees growing in the rocky soils along the clear, greenish waters of the Texas Hill Country.

Steve is an interesting person, and I have found few landscapes that I enjoy more than the rolling hills in his part of Texas. His plot was on a site where the soils are alkaline and chalky, and where rocks abound. The streams are clear, with smallmouth Guadalupe bass, unique populations of cichlids, and at one time in the 1800s, foot-long freshwater shrimp. The streams are lined with extremely tall and old bald cypress trees, believed to be unique ancient populations and part of a forgotten stand of trees that stretched from Arizona into Mexico. Here in this old German-populated part of the state, Steve has collected and cultivated about five thousand crinum bulbs of different varieties.

From Steve's collection I became acquainted with C. 'Stars and Stripes,' a strongly striped variety. It blooms about three times all summer long and is being produced in one-gallon pots by a grower in Houston that I often work with, Heidi Sheesley. It is a newer variety, although discussions about "heirloom" and "non-heirloom" crinum bulbs are tame compared to the highly debated and highly defined introduction dates of daffodils.

Steve and I became acquainted during my internship with the flower-growing operation Hines Nursery the summer before I moved to the red cabin. While working at Hines Nursery, I lived in a hotel room close by. Steve sent me back with a pot of bulbs. I left it on the tile section of the little kitchen area as I rushed to go to work at the nursery. I had seen "NO PETS" signs but not "NO PLANTS" signs, so I figured I was okay.

When I returned from work, an army of fire ants had begun their march from out of their hiding spot in the pot and were surveying new areas of my room. With very few possessions at this point in my life, I did not have any bug spray. Making a quarantine perimeter around the line of marching ants with dish soap, I began squashing them one by one with my finger. After the threat had been neutralized, I wiped up the soap and put the pot in the sink. Most of the ants had fallen in battle, and the rest were not a problem until I could later move the pot out.

Steve and all of the bulb collectors were very helpful, but unfortunately *Crinum* is a genus with mixed identities, labeled under many different colloquial names, so it would take years for my mind to sort out all I had been told. All have Latin names,

Bill Welch with Heidi Sheesley and a Crinum *'Mrs. James Hendry' in the background. CLW*

but distinguishing one plant from the other can be extremely difficult, and they are often sold under the wrong name. However, the diversity of the common names is entertaining and includes names such as milk and wine lily, 12 Apostles, St. Christopher's lily, Queen Emma's lily, and swamp lily.

In most cases, crinums outlive the gardener—on the Southern Bulb Company farm we grow *C. powellii* 'Album,' first brought to Texas by Texas Revolutionary fighters from the South, or so Ben Arcuni was told when he dug it for Southern Bulb. In *Garden Bulbs for the South*, Scott Ogden lists the date of introduction of this bulb as 1888, so perhaps the landowner was a little confused. This particular crinum is one of the most common found in the trade, one of the most cold hardy, and one that can be found in many historic gardens. I have seen it growing in the Chelsea Physic Garden in England, in early eighteenth-century plantation gardens along the East Coast, in an old garden in Maryland, and in the Victorian gardens that dot the state of Texas. There is also a pink variety commonly found in the trade, *C. powellii* 'Roseum.'

I've had the pleasure of staying in numerous Victorian homes throughout my travels. I remember one home in Palestine, Texas, where I can almost guarantee that the clump of crinums has been growing at the foundation of the home since it was built, which makes the point that they last almost forever.

To my astonishment, I still remember the exclamation of one elderly person in an audience to whom I was showing slides of this house: "That's not old; that's a post–Civil War home." It raised a good laugh from me and the crowd. She continued, "My place is pre–Civil War and instead of sleeping in your truck tonight, you can sleep there." I accepted the invitation and had a place to stay!

Victorian home in Palestine, Texas, with clump of crinums growing at the bottom left. CLW

That is how I found myself one early summer evening, driving to stay on an old plantation just outside the city limits of Nacogdoches, Texas. When I was shown my room, we had to shoo away a swift that had taken up residence for the evening. After that little adventure, we took a quick nighttime tour of the premises, and I'll never forget the sweet, comforting smell of the old gardenia bushes in full bloom. The blooms were not the big, overbred creatures we are used to now, but smaller, and they covered the whole bush. With a large number of blooms and the humid summer evening acting upon them to release the fragrance, I was in heaven. The owner, of course, had crinums worked in well throughout her garden, but that is to be expected in these older homes.

Will crinums ever have the same success in modern gardens? Much hope remains for more crinums to become hugely popular in warmer climates, especially when one considers how amaryllis, with very similar blooms, swung into popularity as potted plants. Bloom time, longevity of cut flowers, and use in the garden all need better research and information before a leap into the mass market can be made. I consider it to be one of the top three bulbs with the most potential in warmer climates, but it definitely has some challenges to overcome.

Believing that I could overcome the challenges of marketing crinums, and other bulbs for that matter, I continued to pursue my business endeavors. I drove to San Antonio to meet with Joe Bradbury to discuss the Southern Bulb Company. We had dinner, and as the evening progressed, I told him about the business and the strides we were making. Our conversation was leaning toward a business discussion, and I encouraged it in that direction. I expressed my need of some financial support from somewhere or somebody. Knowing his tough past, I was a little nervous approaching the subject. I continued to tell him about the business and some of my business needs. There I sat, with this crinum of a man, still not sure where this was going to lead.

"Drugs and prostitution, son—the only two easier ways to make money," he repeated once again. "I like you, son. How much do you need?" he asked. All of a sudden he became very sharp and looked at me with clear eyes even though he had enjoyed at least a couple of scotches.

The long, slender stalk of a **Crinum** *'Ellen Bosanquet.' Note the anole, a friendly lizard in the South. CLW*

"Well, I really don't know . . ." I started, but he cut me short.

"Son, how much do you need?" he asked.

"Twenty-five thousand dollars," I responded.

"I'll wire it tomorrow."

I stayed in the guesthouse that night, staring at the brick dome ceiling crafted by masons Joe hired in Mexico, amazed that I would have twenty-five thousand dollars wired to my account the next day. Joe wasn't up when I left the next morning. I drove out of the gated San Antonio community and headed back to my little red cabin in the Piney Woods of East Texas. Leaving Joe's home, I took my business plan that he never read, never even looked at, and had tossed back at me the evening before. The money was wired, and the Southern Bulb Company had a new partner.

CHAPTER 3

Lilies and Bulbs
That Are Called Lilies

S oon pressure increased to grow more bulbs and more varieties. As more people heard about the Bulb Hunter and the Southern Bulb Company, the advice started to roll in, and some of it was adamant that I focus on certain kinds of bulbs. I personally enjoyed all of the bulbs but was faced with the reality that I couldn't collect and grow everything. Struggling to manage a new business with twenty-five thousand dollars, I had to pay some attention to the invoices but try to stay true to the original horticultural intents of starting my business.

Living in the cabin, I had a lot of time at night, alone, to reflect on this. I began to understand that in the eyes of some who were following me and my company, I had a responsibility as a steward of these bulbs. The movement toward heirloom bulbs and preservation of gardening heritage ran much deeper than my company. This was something I understood as well, and I tried to balance this understanding with my needs to keep my doors open. On the whole, Bill Welch offered balanced advice, but he also thought I needed to grow certain bulbs. One of these was the Easter lily (*Lilium longiflorum*).

In our conversations, he often informed me of his own stand of Easter lilies. Over the phone he would remind me of where they came from and how well they were doing and would then ask if I was growing any. "Indeed, they are grand," I would say. "I know. Yes, they are wonderful," I would continue.

"Well, why aren't you growing them?"

A close-up of a tiger lily (Lilium lancifolium). Note the little black bulbils at the base of the leaves in the background. CLW

"I'm trying."

"Where are they?"

"Planted here in the black crates," I said politely as I looked at my soil-filled crates planted with the lilies near the barn at the farm. Bulb crates are plastic containers that are about ten inches deep and twenty-four inches long and very much resemble produce crates like one would find at the grocery store, only deeper. They are made of thick molded plastic and have open slits running across the length of the sides and the bottom. The slits allow for ventilation to prevent disease among the bulbs, and the thick plastic ensures a solid way to stack and handle crates for overseas shipments.

They needed to be transferred to a row in the field, I was informed. I resisted this suggestion for a while but finally acquiesced to moving half the bulbs out of their growing crates by the barn and planting them directly into rows in the main field. Out in the open rows of the farm many things could get to them easier, like armadillos and voles. My hesitancy was justified, I thought. They were, after all, his original stock, from a very special person in Navasota named Pam Puryear, and I did not want to lose them.

Pamela Ashworth Puryear was a purveyor of old plants and one of the original rose rustlers (like a bulb hunter, but focused on collecting antique roses). Many a plant has been named after her, and she introduced a significant number of durable, beautiful roses, perennials, and bulbs suitable for Texas' harsh climate. She lived in Navasota, an hour northwest of Houston.

The climate in that part of the state stays much warmer than mine, though. The clump at Bill's place near Round Top (in Central Texas between Austin and Houston) does not face the harder freezes and longer cold spells that I do at the farm just three and a half hours north.

Bill and his wife Diane purchased a small tract of land, bought a very old house from the turn of the nineteenth century, and moved the house to the recently purchased property. Diane named the house Fragilee because, as I'm sure the reader can guess, it was extremely fragile. "Termites holding hands," as their builder said. After renovations, the result was a truly stunning structure with just as stunning gardens.

Bill and I communicated continually about my business, and I kept him up to date on my travels and bulb-hunting adventures. As our conversations grew in frequency, our friendship also grew, and soon I was staying at Fragilee if my travels took me that way. Eventually, I knew where the spare key was and how to turn up the hot water, operate the radio under the stairs, and wash the sheets before I left. It was my sanctuary in the midst of busy and stressful travel on interstate highways.

I needed to learn, and he needed help working on his cottage and garden, which included everything from deadheading his antique roses and trimming the hedges to painting the window

frames and mowing his pasture. His garden and home were always ready for a magazine to stop by and take photos. I wanted
my brother to meet Bill, so one spring, John and I offered a weekend of service at the cottage while Bill and Diane enjoyed their
weekend retreat and worked on projects of their own.

Each day Diane wrote a letter to a different person, whose address she had written on her calendar. This system allowed her
to maintain and keep connected the eclectic group of friends
gathered throughout their life together. The evening John and
I were working, the Welches had friends over for the evening. I
soon realized Diane possessed a charming gift for making you feel
extremely special.

"This is Chris Wiesinger, the one we told you about who is
starting a bulb business," she said as I was introduced. Soon I
realized that all introductions were given in a way that made you
feel like a million dollars and gave you an air of greatness. They
were delivered with an inflection and look in her eye as if to say,
"You don't know of the great Chris Wiesinger?" Diane could be
talking to the governor's wife and the state's first lady would
soon think you of huge importance. "A Web site is important,"
Bill said in passing but matter-of-factly the next day as we ate the
sandwiches he had made for lunch.

"Bill, they don't have the money for that right now. They have
to eat first, which is much more important than a Web site,"
Diane quipped with an initial drawn-out "Bi-i-i-ll" and glance at
me. I think she was looking for agreement, but politics was essential in this case. I just acknowledged with an agreeable look that
could be taken in support of either person.

"I know, Diane, but out of everything, that's something that's
just gonna have to look good," he concluded with a look my way.
"I'm just saying, if those garden ladies don't like what they see,
they're not going to order."

The conversation continued over lunch and into the dinner
hour. The next morning as I walked down the narrow, steep
wooden stairs, they were sitting together on the front porch
drinking coffee and gazing over the hills dotted with farms and
covered with bluebonnets. The view from Fragilee's front porch
was stunning. The sun had not risen yet, and the grayish light

had only begun to show off the color of the official State of Texas wildflower. By the time I had finished my first cup of coffee, the sun had peeked over the horizon and the blue hues began to turn brighter and brighter. "If Monet only knew," Diane said, and all on the porch agreed but said nothing.

I continued to see the Welches and their friends often. As I traveled throughout the South, they seem to have a friend in every town willing to host me. They also offered their home to me anytime I was in the area, which turned out to be often.

There was a small picket fence of stained wood on a gravel path that led to the back entrance of Fragilee, and on the right of this entrance was a big clump of the Easter lilies Bill was so fond of. They thrived there, which surprised some people, who were used to seeing Easter lilies only grown in pots. This patch of perennial Easter lilies makes sense, however, because Easter lilies are grown in more tropical environments. They actually *like* our warmer climates and were once produced commercially in the South. Now production primarily exists in Northern California and Oregon, and the bulbs are precooled and forced to bloom in time for Easter. Without being precooled and grown

Easter lilies blooming right on schedule at St. Andrews Episcopal Church in Houston Heights. The church has been there for more than one hundred years. CLW

at certain temperatures, Easter lilies in the South bloom closer to Mother's Day.

In college, when I was called "Flower" in the Corps of Cadets, my horticulture degree focused on forcing flowers like Easter lilies to bloom. I worked in the university greenhouses for Terri Starman. Lily growers routinely sent new varieties of bulbs to researchers at academic institutions, and Starman was one of their favorites. By using the right techniques, she could make most any lily bloom on schedule and still have compact, good-looking flowers and a standardized appearance. Her expertise allowed outlets such as grocery stores or retail garden centers to have uniformity in their plant displays.

Once the bulbs arrived at the greenhouses, Starman's research assistant placed them in the cooler, sometimes for chilling requirements but mostly to keep them dormant. We then put them in individual pots and forced them to bloom. Forcing plants to bloom involves different processes of drying a bulb out, replanting it, watering, and controlling the time the bulb is exposed to different temperatures and light. Starman's forcing regime strictly followed long-established guidelines to ensure the best possible results. Forcing is not normally a process successfully achieved by home gardeners.

Commercial Easter lilies can actually be planted in the ground after they bloom in their pots. Once growing in the ground, they won't usually bloom around Easter but, as noted earlier, around Mother's Day. All over the South they can be seen blooming in early to mid-May. The Heights area of Houston, one of the first neighborhoods in the city, has gone from very historic to very trendy. Although developers have torn down some structures and replaced them with modern homes, some sections of the neighborhood have been preserved. One such place is the Episcopal Church, where a very nice stand of Easter lilies blooms every year.

I was still struggling with Easter lilies on the farm—my crated growing system had not worked for them. A couple of weeks after I followed Bill's suggestion and we moved the testy bulbs from my black crates to the rows, an armadillo dug them up for dinner. I guess in the end we both lost the argument about how to grow them, but the bulbs were the real losers. I have since given my remaining bulbs to specialty growers in hopes that we will someday be able to grow them in larger numbers. Another white lily that blooms reliably in the South but at a completely different time is the Formosa lily or Philippine lily (*L. formosanum*). A favorite among gardeners, it is reminiscent of ghosts dancing in the garden as the wind blows the pale white blooms on long stalks. A seven-foot-tall stem can really catch the wind. The flowers are long, white trumpets with maroon midribs running down into the throat of the bloom. The visibility and intensity of this maroon stripe depend greatly on the origins of the first collected bulbs.

High degrees of variability also exist in how tall the bulb will grow, and some dwarf varieties are available. One such smaller variety grows at the MD Anderson Cancer Center in Houston below one of the sky bridges and along a walk. Native grasses are tucked along the base of the plants, and the mix of grasses and lilies is stunning. Head gardener Joseph Johnson gave Bill and me a tour one day and showed us how he had created this combination. He even later collected copious amounts of seeds and sent them our way. Joseph is now at Shangri La gardens in Orange, a town in southeast Texas near the Louisiana border.

We were at MD Anderson for a sad reason. Shortly after I started the bulb business, doctors diagnosed Diane with melanoma. Our hearts were heavy as we visited her while she was undergoing treatments. Nine short months after being diagnosed, she passed away. Not before, however, she had a chance to see her granddaughter baptized and to have time with the new baby girl who crawled around on her bed. I had experienced a similar situation with the death of my mom nearly a year and a half earlier. It was a strange time for me—to be in the red cabin, starting a bulb farm, and have so many changes happening in my life.

During this time, John was living in Atlanta. Once settled, he kept noticing a tall, large, orange-flowered lily commonly called a tiger lily (*L. lancifolium*). John quickly informed me that I needed this one. I agreed. Everyone would want the tiger lily. Tiger lilies can be found mostly just north of the coastal areas, and many people talk about seeing them all the way up into Missouri, where they can thrive in ditches.

The bright orange of the tiger lily will catch your eye, even though it will most likely be growing in a shady garden. Three- to four-foot stalks are topped with a rich, bright orange flower covered in black specks. The petals furl open and allow wide sections of the flower to make a high visual impact. The flowers will eventually point down, but this does not diminish the blooms' impact.

To best illustrate how tough these heirloom bulbs are and how easy they are to propagate, I tell the story of a lady with a yard full of tiger lilies. Sadly, I cannot recall her name. When she was a girl in Missouri, she longed for a tiger lily from an old man's yard that she walked by every day on her way to school. After she had asked again and again, the man one day conceded and gave her one, just one, bulb. Seventy years ago, when her family moved to Texas, she carried that tiger lily bulb with her. Since then, that one bulb has filled the bases of trees, shady flower beds, and the foundations of her house.

Bill hasn't had luck with them in College Station, but he has grown tiger lilies at his home in Mangham, Louisiana, a 1905–6 late Victorian house with longleaf pine floors, cypress siding, eleven-foot ceilings, and a moderate amount of gingerbread trim. The garden is tended by Jessie Lee Harris, a longtime helper of the Welch family. When I stay there, I most often come in late and leave early. Jessie Lee leaves the key for me, usually with some homemade goodie in the kitchen prepared by Jessie Lee's wife, Mildred. Sometimes I stay for breakfast, and one thing I don't do is leave without a few plants from the garden. Then it's back on the road looking for bulbs.

Tiger lilies bloom at Mangham in July. Bill isn't sure how they got there, but the lady who owned the house for many years was an excellent gardener. I have walked around from plant to plant collecting the little black bulbils that form at the base of every leaf. These bulbils are interesting and appear to me as miniature black bulbs. I find them by spotting the bright blooms of tiger lilies as I drive through southern towns.

My ability to drive and look for bulbs is getting better and better. Driving with a plant hunter can be dangerous. Be prepared for the brakes to slam and for the car to make a U-turn. But that's not the most dangerous part. The specimen that catches my eye is often located at a private residence exhibiting all the signs of

Harvesting the seeds from a tiger lily (Lilium lancifolium) *deep in the heart of East Texas. CLW*

"stay away." By this point, my standard opening line is getting better and better as well.

With the car parked and adrenalin running, I step out of the car, walk up to the front door, and knock. Overcoming the fear of knocking is the hardest part for me. When the door opens, I tell the person who I am and what I'm doing, then plainly lay out a proposition to acquire some plants from the yard. It usually sounds something like this:

"Hello, my name is Chris Wiesinger, and I make a living collecting plants from around the South. One of the plants I've been looking for is in your yard, and I was hoping I might be able to trade you some plants for it or pay you for some."

"Really, what kind of plant is it?" the person usually asks, now stepping out from behind the screen door and leaning out to see the plant.

"Well, it's a _____, an old-fashioned _____." I fill in the blanks accordingly.

"Sure, you can have some. Just fill the holes back in once you are done."

"I'll do that for sure, thank you!" And that concludes the dialogue. Walk back to the truck, grab a shovel, collect some plants, leave some behind, fill in the holes, and go.

The first time I collected the bulbils of the tiger lily, I knocked on the door of a house outside Henderson, Texas. A really big man answered the door. He was tall, muscular, and a little bit rough around the edges. I gulped and continued with my familiar story. His response was cordial, gentle, and friendly, and he knew exactly what I wanted and would help me collect them. As we walked around the garden from tiger lily to tiger lily, he told me about his time in the Special Forces as a Navy Seal. Mixed in with the wartime stories, he gave me advice about the bulbils. "They will rot on you!" he exclaimed. "You just can't leave them in a coffee can. I did it once and came back toward the end of summer and they were all rotted."

The bulbils collected with my Navy Seal friend are now large, flowering tiger lilies, each producing blooms and ten to twenty new bulbils every year. Tiger lilies do not enjoy growing in full sun, but with moisture and shade, they really take off.

My truck has only two cup holders, in the middle seat, one for me and one for the random small number of bulbs or seeds I always seem to have with me. Originally it was my brother's truck, and he usually names his vehicles, but this one simply remained "the blue truck." The transition of ownership from my brother to me was gradual and was eventually decided when the stick-shift, V-6, step-side pickup approached one hundred thousand miles of bulb-hunting activities. With coffee in one holder, and an empty coffee cup filled with little black bulbils in the other, I made the short jaunt up to I-20 from Mangham in the blue truck. From there it is about a three and one-half hour drive back to the farm. Along this highway, as well as I-45, I-30, I-95, and so many others, I often peer out the window, looking at all of the bulbs left behind at old homesites destroyed to make way for the highways connecting one coast to the other. Having been down the highways from one end to the other, I often marvel at the change in plant life across such an expanse. The tiger lilies are beautiful but are not natives. There are, however, some lilies found in the ditches of these roadsides that are considered natives.

These are bulbs that help paint the scene of Texas horticulture for me. One such bulb is called a "ditch lily," although it really isn't a lily. It is not even in the genus *Lilium*. Around mid-April, white spidery blooms appear in the marshlands and bar ditches across the South. It is one of our truly spectacular native bulbs, and we have records of it from some of the earliest expeditions. When John-Louis Berlandier, a French naturalist, led a Mexican scientific expedition across the state of Texas one spring in the mid-1800s, he recorded walking in a field of knee-high white lily blooms. It is believed he was speaking of *Hymenocallis liriosme*, the white spider lily, or as Frank McMains, a Louisiana reporter said, "They will always be ditch lilies to me."

I spoke to the local Master Gardener group in Baton Rouge one afternoon, and he joined us for dinner. His stories seemed more entertaining to me than my own. On one occasion, authorities allowed him to write about the boxing matches at a high-security prison in Angola. Frank's adventures included looking for ferns by an abandoned plantation a couple hours up a Louisiana river and hosting traveling bands as they passed through on their way to Austin for a music festival. He also owned a bar in downtown Baton Rouge that catered to a quieter academic crowd. Frank wanted to go on a bulb hunt with me, and the idea sounded great to me. I needed to find more places where *Hymenocallis* grew, so we embarked on a search for ditch lilies.

In gardens, we are more likely to find the improved cultivated selection *H.* 'Tropical Giant.' It survives better in garden conditions than its native brother, *H. liriosme*. It has long, large, glossy leaves and blooms in July. This 'Tropical Giant' multiplies and blooms very quickly and can be seen in gardens in the northern parts of the middle South (zone 8a) and in coastal areas. Remembering my trip to La Salle, Texas, I know this particular bulb requires a special digging tool, as it also can make large, solid clumps.

On our spring trip to find the native *H. liriosme*, I figured on exploring roads that would take us to Monroe, Louisiana, allow-

A clump of Hymenocallis by a crab apple growing in John's backyard in Tyler. These plants are often called "white spider lilies" close to the Gulf Coast. CLW

ing a visit with my Monroe Garden Club friends along the way. There is one firecracker of a woman who lives there and who has taken me on the most interesting bulb hunts around the area—I consider Adele Ransom to be one of my favorite people I have met on my journeys. This time she insisted we meet a friend of hers close to the Arkansas border with a garden we had to see. I decided that Frank might enjoy this adventure. First, I checked my calendar, to make sure it wasn't liver Thursday. (She had liver every Thursday for lunch, and one of my earlier visits was ill timed for my stomach.) The date for the hunt was on a Wednesday, so everything looked fine.

Frank met me in Mangham, and we left for Monroe for the garden Adele had recommended. The garden was expansive and beautiful, although still in its earliest stages. The gardener was deaf, which influenced the architecture of the home and led to many windows that looked out over the grounds for a more visual impact. On a side note, she also had a male ostrich, which

A fellow bulb hunter, Eric Breed, and I go looking for the native Hymenocallis liriosme *in swamps. CLW*

[Image of a man with a camera looking at native Hymenocallis liriosme in a swampy field]

happened to be in season and kept charging us. We doubted whether or not the fence could contain this extremely large, aggressive bird! After the garden tour, Frank and I began looking for the *Hymenocallis*.

Many abandoned roads and a few Confederate flags later, we hadn't found anything. Not all bulb-hunting adventures are profitable, and we struck out that day. I invited Frank to dinner in Monroe with a few of the members of the garden club, but he had to head back home. A couple weeks later Frank sent me the most amazing picture of a field of *Hymenocallis* he had come upon. We are planning on more adventures together down the road, and I am glad one more person is excited about bulbs!

After our failed hunt, I stumbled upon a large patch just north of Houston. Another bulb collector from Holland, Eric Breed, was visiting me at the time and had come to Texas to see the farm, hear me talk, and stay at the cabin. He had an equal appreciation for the bulbs, so when I put the brakes on and spun the truck around, he eagerly followed suit, bumping along in his rental car. White blooms filled a swampland area just south of Magnolia, Texas. Barbed wire cut through the patch of blooms, making a fence between the roadside ditch and the private property. Woodlands merged with the swamp about twenty feet into the private property, and where the woods began, the lilies stopped. It was a typical, swampy, *H. liriosme* scene.

The more native *Hymenocallis* continue blooming on into the summer, along with the other native *Hymenocallis* species that bloom then, such as the Cahaba lily (*H. coronaria*) in Alabama, also called the "Shoals lily" in Georgia and South Carolina. It actually lives in some of the rivers, in the shoals or shallow areas. The shoals also often mark the transition of two different geographic regions in the area, the Piedmont and the Coastal Plains of North Carolina, South Carolina, and Alabama. Ships navigating the rivers were forced to use canals to go around these shoals. The bulb blooms in late May to early June. There are several festivals dedicated to it, one of the most famous at Landsford State Park in South Carolina.

Although a native, just like the *H. liriosme*, it is often compared to the nonnative *H.* 'Tropical Giant' selection that is the most popular for the garden setting. It can also be hard to find at times but is more readily available commercially. At the Southern Bulb Company, we have found that customers respond well to the larger bulbs of the 'Tropical Giant' and also like the lush green foliage that remains on the plants each summer.

Grant Cox joined me on the trail looking for large masses of *H.* 'Tropical Giant' around the central part of East Texas, primarily around the historic town of Crockett. Grant, then a college student at Texas A&M, was working his first summer for us and later spent one amazing summer on a sales tour with the Southern Bulb Company. Grant's dad once practiced dentistry in Crockett, and this connection reminded me how small the state actually is. Grant and I went up and down every road, marking each location of bulbs with a primitive GPS device. We were excited about our finds but never actually returned to dig any bulbs. After a long day on the road, we returned to the cabin to crash with the rest of the growing crew.

A close-up of the Hymenocallis bloom. CLW

The "crew" at the Southern Bulb Company continued to enlarge because our one investor, Joe Bradbury, had put additional funds into the company. These are some of the crew of workers and friends; I will most certainly have to leave some out, but they all had a very important role in my life. Ben Arcuni, a lion of a man, was my former first sergeant when I was in the Corps and later transitioned into a flower lover and worker/part owner with Southern Bulb. Brad Gaultney was a good friend and employee and eventually part owner of the company. Jacob and Hayley Shalley brought my dog Fischer into my life, and Jacob helped with construction projects on the farm. Chad and Sarah Jones were a continual hotel to me in Mobile, Alabama, and Chad went on some of the earliest bulb-hunting adventures with me. You will continue to hear these names throughout the story.

In 2006, two years after starting my business, my bulb-hunting adventures continued as Brad and Ben "held down the fort," allowing me to travel and grow the business.

Ben found and dug our selection of *Crinum powellii* 'Album.' Ben came to work for the Southern Bulb Company from College Station. When I met him as a freshman, it seemed like he did nothing but keep us moving, and moving fast. I knew that he would be up for the tough, dirty life on the farm and on the road. His eye quickly adjusted to finding good bulbs and developing relationships with people. Ben's discovery of this particularly nice selection of *C. powellii* gave us something that is very much enjoyed on the farm every summer.

Brad was also a huge asset to the company during this time. He left a high-profile, fast-track-to-the-top General Electric job to come work for the company. When I first met Brad, he was nothing more than the other bald-headed cadet across the hall that stood at attention next to his door as I did next to mine. Ben was in charge of us both. Brad would make faces when the upperclassmen weren't looking. Attempting to make me smile was part of his twisted sense of humor, I suppose. It bothered me because I knew if the upperclassmen caught us, we would most definitely pay for the offense. Giving in to peer pressure, I would normally return one very quick signal with my eyes. Brad was in better shape than I was and could handle the impending punishment, but it definitely would not be fun for me.

From left to right: Ben Arcuni, Zac Coventry, Brad Gaultney, and Grant Cox standing next to a patch of parrot gladiolus (Gladiolus dalenii). CLW

GE Brad, as we jokingly called him, had great business sense. When GE Brad made a suggestion, it was wise to listen. He is also a great salesman. One summer while in college, Brad sold books door to door and was the number-two bookseller in the nation for that particular company. At General Electric, in its supply company, he managed large corporate accounts and regained business in areas where they had lost it. He always offered advice for my upcoming sales meetings with words to say, words not to say, and body movements that would help close a deal. It did not come naturally to me, but Brad helped me through it and became an integral part of the business.

Other friends helped with my business as well, some of them just on the weekends, to help with construction and farming or just to give encouragement. Visitors are always welcome at the cabin. One of my closest friends, Jacob, a construction science major, helped with building projects on the farm. He came up one weekend to the cabin with his wife, Hayley.

"What are we doing for dinner?" Hayley asked. "Fish," I said, "but we have to catch them." Hayley's big eyes grew even bigger, and she became excited, picked up a rod, went out, and caught a fish. Jacob, never to be outdone, caught the rest of dinner. Later, as I traveled the state, I would often end up on their sofa in Dallas, or Houston, or at his parents' house along the coast. I mention their willingness to take me in and house me because it is representative of the hospitality that so many showed me. This was not just in Texas; I had sofas to stay on from Louisiana to Washington, D.C., down to Florida and all the way to California. It was kind of like being a vagabond, but with a specific purpose in mind for my travels.

I also mention Jacob and Hayley because they were a big reason the Christmas party at the cabin became an annual tradition. After I had lived in the cabin for a couple of years, they rounded up support from all of our friends and bought me a present. They decided that it would be best for my life if I had a dog. Selecting the perfect breed for me, they purchased a Weimaraner, a German hunting dog. So, at one Christmas party, König Fischer (King Fisher) came into my life.

At first, Fischer would sleep a lot in my truck or in my office on the bulb farm. Eventually, he grew bigger and bigger. I took him on my crinum-hunting adventure late the next spring to La Salle (where the crinum-digging tool was invented). The ladies in La Salle adored him but warned me about letting him roam too close to the canals and too far away from me for fear of alligators. He would normally stay very close to me, sticking his nose in the holes I would dig around crinums. Never actually helping dig

Fischer comes into my life. Jacob and Hayley with me after they gave him to me at the annual red cabin Christmas party. CLW

Fischer wasn't really sure how to help, so he would just sit right in the middle of wherever we were working. CLW

a crinum, he just provided entertainment, and I have some great pictures of his big nose smelling a crinum bloom.

With the support of friends and the increased support from investors, we set up a permanent location for the company on a piece of land separate from the cabin. Brad and Ben worked with me hand in hand as we purchased a piece of land and began to set up the bulb farm. We built a small pole barn to house a newly purchased tractor, drilled a water well, brought out power, brought in an office trailer, and ordered a phone/fax line that also had Internet. On the property were two nice hay meadows, separated by a large ravine filled with trees. At the entrance to the property were two not-so-nice hay meadows. That is where we put in makeshift roads and the temporary office. The nice hay meadow in the back behind the ravine we turned into our new farm plot. Inside the trailer we began to work on the company's first Web site.

Much to Bill's delight, GE Brad managed to get a Web site up and running. Web sites have been some of the most painful things we have had to work through, but they have been worth it. In the spring of the next year we were mentioned in *House and Garden* magazine, and our Web presence no doubt helped. The magazine included me as a "Top 50 Tastemaker" for the nation, and this would involve a trip to New York. The man responsible for this was Tom Christopher, an avid garden writer and contributor to the magazine. He asked Bill if anything young or exciting was happening in the plant world in Texas, and Bill gave him my name. One thing led to another, and I was on my way to New York City!

The trip to New York was a success, and in a roundabout way somebody from the *New York Times* heard about what we were doing in Texas. Our jaws hit the cabin floor when the phone call came in. A reporter was actually going to fly down and spend several days looking for bulbs with me. I picked her up at the Dallas/ Fort Worth airport during what had to be one of the hottest weeks of the year. She hopped in my truck, and we were off for East Texas. Reserving a room at the local bed and breakfast, she stayed just up the road from the cabin. Her visit occurred at a time when few bulbs were blooming. The tiger lilies had already finished, the Philippine lilies had not started, no crinums were in bloom—it was a transition week during the summer. The next week lilies and crinums would be in bloom.

The long line of thunderclouds with quickly cooling temperatures and lightning bolts from the week before had long given way to the hot, humid days of summer. Plants were melting. I was melting. She was melting. Warm golden rays of sun poured in through the cab of my pickup truck, and a little fairy must have slipped in and sprinkled sleeping dust over us. Perhaps, if I had been a better conversationalist, it could have helped the mood, but there was a very professional reporter/subject of story relationship. What I really needed to do was find a blooming bulb, and fast.

In Lufkin, Texas, home of the infamous congressman Charlie Wilson, we found the answer. A line of thundershowers had come through, offering a quick change to the weather: slightly cooling temperatures, moisture (from the rain), and nitrogen-charged particles (lightning in the air). Nobody knows for sure how much, how little, or what combination it is, but these are sure-fire conditions to coax a rain lily into bloom. As we slowly drove the back alleys and run-down areas in Lufkin, in one yard along a chain-link fence (hurricane fence as they are called here) was a little pink bloom tucked back in the shade. It was the bloom of a rain lily.

With a deep breath and smile, I put the truck in reverse and hopped out. The reporter sprang into action. My hand was barely on my door handle before her door popped open, and she jumped out. I was a little shocked because normally someone is a little confused about the situation and asks, "Should I get out?" Confused or not, she was going to be a part of the action. Together we bent down and peered at a pink rain lily.

Similar events happened a little later in the day except with an amaryllis and crinum discovery. I quickly hopped out to find the owner of this piece of property. The owner ran the Ford dealership, a couple of lots over, and the reporter and I briskly walked over to seek him out. Because the reporter and I entered the floor of the dealership simultaneously, the sales representative must have thought we were in competition with each other. This reporter was not going to miss any of the action and recognized that this was the story she was trying to capture.

With permission received, we walked back to the lot and I pulled a couple of shovels out. The amaryllis were going to be easy, but the crinums had been long established at this 1940s home. Warnings were issued about the tough bulb, and I transferred the shovel into her hands. The bulb remained in the ground, and my worry and concern must have been evident because she eventually just handed the shovel back to me. Doing what I did best, I dug. Up popped the crinums, along with smaller dormant bulbs that turned out to be spider lilies (*Lycoris radiata*). She could now mention in her story the dormant bulbs and flowering bulbs, with several finds on this quick adventure.

On the drive back home, the same sleep fairy entered the truck and caused the reporter to doze, but we were soon at the bed and breakfast. Brad, Ben, and I washed up at the cabin and prepared to meet her for dinner that evening. First, I wanted to take her

Bill and Diane Welch with Ben reading the New York Times *article CLW*

fishing on the lake, especially since she had never been fishing before. We hopped in the boat, caught a bass and some bluegill, and she left the next day. The reporter called from New York before the story came out. "I'm going to call you something different," she said, "and it's kind of a big spread, so I don't want you to be surprised."

The next day the two-page color spread came out, and in big bold letters the title read "The Bulb Hunter." I had gone from "Flower" to "Bulb Boy" to "The Bulb Hunter." The story was picked up in several newspapers around the nation, and that was the beginning of the name and many more adventures that would make up my quest.

I suppose after all of this, I'll get to the real botanical point of this chapter. What makes up the name "lily"? The rain lily that the reporter and I found on our bulb-hunting experience was not actually a lily, just as the *Hymenocallis* is not actually a lily. An interesting thing to note about lilies is the common name "lily" seems to be applied to any flower with a trumpet-shaped bloom (or in some cases, people will call anything a lily): cemetery lilies, swamp lilies, ditch lilies, crinum lilies, and rain lilies. Rain lilies are in the amaryllis family, and they belong to several different genera of plants, including *Zephyranthes*, *Habranthus*, and *Cooperia*. But why should the name "lily" not be used for this special bulb? Lily implies something sweet, beautiful, and intoxicating. Rain conjures up fun, dramatic, and refreshing images. What happens when you combine the two? The answer must be romantic love, of course, and I had a meeting coming my way that I did not expect.

CHAPTER 4

Rain Lilies

B y now you might have realized that much of my story is not just about the bulbs but about the people. Fast-forward several years into the business. Before I knew it, garden centers, clubs, and Master Gardener groups requested me to speak to their respective organizations. The Arbor Gate, one of those garden centers, holds a special place in my heart.

Every spring and fall I speak at the Arbor Gate in Tomball, Texas. On each occasion, I see white spider lilies (*Hymenocallis*),

A clump of pink rain lilies (Zephyranthes grandiflora) *is welcome beauty in the middle of our hot summers. CLW*

a longtime resident of many old gardens and native settings in that area, blooming in the fields on my trip down from East Texas. The Arbor Gate has an open-air stage for presentations, and instead of my usual PowerPoint presentation, I talk for about forty-five minutes outside with only plants in hand. After one fall talk, an attractive girl who worked there boldly approached me and asked if I was dating anyone. A little shocked, I said no, and Amy asked if I would be interested in meeting her sister.

A million doubts ran through my head in a split second, and I took the easy road out. I responded by saying the standard, "Thanks, honored, but not right now, I'm really busy." A winter passed, and it was time for my presentation at the Arbor Gate in April. After my talk I went to the nursery break room, and there was Amy. Sitting down to enjoy the lunch prepared for me, I was content, happy with life, didn't need a thing, and then my mouth opened and out of nowhere came, "Amy, why not?" I spouted, as if months hadn't passed since our last conversation.

She turned around and, with a sly grin turning into an all-out smile, drew out a long, "Really?"

I replied, "Yes, really," and smiled awkwardly.

Through a series of conversations, delays, and other excuses for not meeting, Rebecca and I finally had a blind date set up in Huntsville, Texas, where she was living. Huntsville is a quaint college town that was for a time the home of Sam Houston and now is home to the university named after him, Sam Houston State University. It is also home to one of the most infamous prisons in the United States.

Coming into Huntsville from the north, you will be invited to visit the prison museum, which is advertised by a big billboard showing a picture of an electric chair. Arriving from the south, you will be greeted by our famed former governor and president of the Republic of Texas, Sam Houston himself: a sixty-seven-foot, white concrete statue.

I met Rebecca on a blind date at the Café Texan in downtown Huntsville. She was wearing blue jeans and a green floral top when I spotted her from my seat, through the greasy glass window of the restaurant, across the street at the corner of the courthouse lawn. She looked left, then right as she neared the road. It hit me that I wasn't just stopping in for lunch; I was on a blind date, and this stunningly beautiful girl was coming straight to my table. Breathe, smile, look natural, swallow, arrange your silverware, breathe, fidget with silverware, breathe. "How had I managed to get into this position?" I asked myself.

I made it through the blind date. In fact, I probably broke some dating rule by calling and asking to see her the next day. Multiple dates followed, and then the real test came. I couldn't make a four-and-a-half-hour drive almost weekly just to see her. At least that is what I told myself, because the gas and time were too expensive. However, I looked at my to-do list and saw that I had plenty of "business" to take care of in that area. I asked if she would spend Mondays with me during that summer, promising to make the drive each week.

"Mondays could sort of be our days," I stammered. She replied with an excited, "Really, really, really??" It was settled. Our Monday dates consisted of bulb deliveries to nurseries, meetings with clients, and of course, bulb digs. Rather than start with a horrific crinum excavation, I decided to break her in gently by beginning with a nice, easy afternoon of seed collecting.

I had to collect seeds from a pink rain lily (*Habranthus robustus*) that was blooming in College Station and asked if she wanted to join me.

I prepped Bill about the upcoming "seed collection" and possible visitor. By this point, I knew where he had hidden the key to his College Station home, and whenever I was close to College Station, I would stop by. The laundry room always had soap, and the fridge always contained something to eat. This time, though, he was going to have two visitors, and he was delighted to have the opportunity to chaperone two twenty-somethings at his house. Chaperone he did.

Rebecca and I arrived in College Station from two different directions, but in time for dinner. Bill told her many things about me that were absolutely not true, but at that time I think most of what was said went in one ear and out the other. We weren't there to talk about me; we were there to collect rain lilies.

I had been intimately acquainted with a particular patch of *H. robustus* belonging to Cynthia Mueller, an avid plantsperson

Rebecca Joy reaching for the seeds of a pink rain lily (Habranthus robustus). CLW

from College Station and friend of Bill Welch, for some time. When the *New York Times* came to interview me and I was scouring the state looking for bulbs in bloom that hot week, I asked Cynthia if I could bring the reporter by her garden. She had a better bulb collection than most people in the state, and there was always something blooming.

Unfortunately, when the reporter arrived, she wanted to do a bulb-searching adventure out into the unknown with no sure finds. This added plenty of stress to my life, but it kept it authentic. I did not forget about Cynthia's bulbs, though. Her small patch could supply all I needed in the way of rain lilies.

The first time I encountered *H. robustus* was in Calvert, just an hour north of College Station, while taking a walk one summer evening with my friend Natalie Anderson. Her house in Calvert was a 1930s bungalow, an arts-and-crafts-era structure that was considered modern in this old town filled with Victorian houses.

Walking the uncurbed streets, some of which were dirt, Natalie and I came upon the house of one of her gardening friends. There in the front bed was an *H. robustus* with a very large, trumpet-shaped, pastel pink bloom only a few inches off the ground. It seemed like an amazingly large bloom for such a small plant and so close to the ground. Later, I realized that these blooms, and the bulb's ability to produce masses of such blooms, explained why it was such a prized possession in hot, humid gardens.

Habranthus robustus produces many papery seeds after each bloom matures. As the flower fades, a seedpod forms in its place and eventually bursts open naturally, allowing the seeds to be spread by the wind. Because the bulbs bloom four to six times from summer through fall, the result is bags of seeds ready to be planted. Once planted, they can grow to a blooming-size bulb within one year. Bulbs like this that can reproduce so successfully from seeds have the ability to naturalize in a landscape.

I use the word *naturalize* rather than *perennial.* When a bulb naturalizes, it becomes thoroughly established as if it belongs there although it is originally from a different area. This success is achieved by bulbs spreading by seeds and other mechanisms of propagation such as division. Generally, all bulbs that naturalize are also perennials. Perennials are nonwoody plants that survive and bloom for two or more years. Some bulbs may be perennial, but that does not mean that they will naturalize.

As an example, we can compare *H. robustus* with another pink rain lily, perhaps the most common and traditional, *Zephyranthes grandiflora*, which I will refer to as the grandiflora rain lily. It is definitely grand—one of the largest-flowering types of rain lilies. Its blooms open more fully than those of *H. robustus* and look almost like stars whose tips are widely rounded and slightly curved back.

Grandiflora rain lilies do not set seeds, so they will not generally spread over large areas. But they return every year and eventually make nice clumps of bulbs. Therefore, we consider them perennials, but they will not naturalize. In contrast, I have seen a home landscape in Houston where the *H. robustus* bulbs have spread so much that they cover the lawn and look like a native planting on an undisturbed Texas prairie. This is something the grandiflora rain lily would not do.

Finding bulbs with the New York Times *photographer at an 1800s homesite in Central Texas. CLW*

One native Texas rain lily that does naturalize is sometimes called the "giant prairie lily," and it excited some of the early botanists who explored Texas in the 1800s. It blooms sporadically throughout the summer, and if rainfall is adequate, it has been known to bloom prolifically three times from early to late summer. Its botanical name is *Z. drummondii*, after Thomas Drummond, the botanist who introduced it and also discovered species of phlox and other plants.

When a nonprofit organization in Fort Worth decided to plant native rain lilies along the Trinity River, they contacted me as their supplier. Streams and Valleys, Inc., directed by Adelaide Leavens, wanted to beautify the artificial levies and surrounding areas that controlled this stream (prone to flooding) as it flowed through Fort Worth before making its way to Dallas. This was a welcome project for me because we could make the bulbs more known to the general public and engage young professionals and groups such as the Boy Scouts to help plant the bulbs.

When Adelaide called and said that they needed anywhere from six to eight thousand bulbs, I thought, "Wow, that is a lot, but not a problem." I called our source in Central Texas. He didn't have that many, and even those he did have were buried a foot deep in dry, hard soil. I mentioned that I thought they clumped, and he assured me that while some clump, many of them grow as individuals.

Thankfully, *Z. drummondii* produces many seeds, which are viable and, although a little slow, can be grown in containers in a greenhouse. The bulb itself does not grow well in garden settings and seems to like being able to dry out during the summer months. In this case, we scheduled a collection of seeds during the first, most prolific bloom in the spring and arranged for the large number needed to be grown in flats.

Other rain lilies have high potential for success in the future, especially in a garden setting. *Zephyranthes* × 'Grandjax' is a pink flower that has continued to impress me. On the edge of the cornfield on Bill Welch's property thirty minutes southeast of Monroe are rows of bulbs he has planted over the past twenty years. Unfortunately, not all of them have done well. A little drift here and there of aerial herbicide applications and some dry summers have been hard on the bulbs and hard to fix while living out of town. One midsummer I tromped through the edges of the corn patch, letting my hands run through the green stalks as I enjoyed the cooling evening breeze. I came upon an abandoned row of bulbs and to my surprise saw rows of pink blooms bursting forth! They were *Z.* × 'Grandjax.'

Bill had shared some of these bulbs with me once, dug from his country home near Round Top, Texas, with strict orders that I be sure to propagate them. I sent some off to a greenhouse grower, planted some on the farm, and mailed the bulk for safekeeping to my brother's house in Tyler. He combined them with another perennial ground cover in pots by his front entrance. The ground cover was a grayish-green succulent commonly known as "ghost plant" or "hens and chicks" (*Graptopetalum paraguayense*).

Squirrels dug up the rain lilies a little later that afternoon. The pests did not remove the bulbs, just dug them up and simply left them on the surface. John replanted them. The volley between

A clump of Zephyranthes **X** *'Grandjax.'CLW*

the two continued until the squirrels finally left the bulbs alone long enough for them to take root. Once that happened, the squirrels didn't touch them. The lilies were pretty that summer and into the fall, blooming in the pots off and on several times depending on the rains. However, bulbs are seasonal, and they enjoy a short winter's rest. They are not dead, just dormant, and will return the next year bigger and better.

That winter I made a quick visit to John's house and saw pansies in the pots. "Oh, pansies on top of the bulbs is a great idea for a winter color change!"

"We pulled the bulbs out," John replied.

My heart sank. Had he lost all sense of the value of bulbs to have just thrown them away for replacement by mere, cheap, modern pansies that were going to die and not resprout?

"Where are they?" I gasped.

"I'll check."

It turned out they were by the trash can; it was almost traumatizing to see them tied up in black trash bags, ready to be thrown out with the refuse.

"If Bill knew this had happened to his rain lilies, he would have nearly killed me," I thought to myself. I took the bags back to the farm and opened them up. To my great surprise, they had all responded to this moment of stress by bursting forth in a beautiful pink show of color.

This strange series of events caused the bulbs to bloom prolifically while in the trash bags. Perhaps it was the imminent death they felt that made them bloom in one last-ditch effort to survive. Or perhaps it was their drying out and being watered again that caused this bloom. Understanding this process better will allow the bulbs to be forced to bloom for a commercial market. Until that is figured out, I am happy to say they have continued to bloom and do well on the farm!

I have rain lilies dotted around the farm, and some tweaking of locations was needed for our white rain lily (*Z. candida*). The white rain lily is quite common now, found breaking pots at various nurseries across the nation. After I gave a talk to a literary club in Tyler, one woman ushered me to her side yard and showed me a line of these white rain lilies in front of her pansies and roses. She

After being thrown away in a black trash bag and replaced by pansies, the rain lilies were in full bloom when I rescued them and opened the bag! CLW

mentioned throwing some away because she did not have anyone to give them to. Cue the Bulb Hunter theme music as I change into my uniform and show up on the scene with shovel in hand. "About to throw bulbs away" is generally my call for rescue.

We had some trouble with our production of *Z. candida* on the farm until I saw something on a trip that taught me a lesson. I had the pleasure of visiting Longwood Gardens in Pennsylvania one fall as part of a group that was treated to dinner and dancing in the gardens' nineteenth-century conservatory. The magical evening continued as we stepped out of the conservatory and onto the terrace to view the fireworks set to the song "Waltz of the Flowers" by Tchaikovsky. Upon leaving the conservatory, I noticed familiar *Z. candida* planted in the bog garden. They were planted in pots with the entire pot completely below the surface of the water. My suffering bulbs were being grown in some of the driest areas on the farm.

To solve this problem on the farm, I filled black bulb crates with dirt, planted the bulbs, and dropped the crates into water. (See the rain lilies section in "A Primer" in this book for a more detailed explanation.) The bulbs look amazing now, but the black crates that adorn the edge of the water and grass now make nice homes for the many water snakes that live at the Southern Bulb Company.

The fact that *Z. candida* likes water made for a great PBS segment we filmed in Rhode Island, although it didn't start off too well. For the show, "Cultivating Life," hosted by Sean Conway, I spent three hundred dollars to ship rain lilies and Philippine lilies, only to have all of the plants overturned, stalks broken, and dirt scattered everywhere on arrival. The *Z. candida* survived though, and we were able to plant its four-inch black plastic pot in a separate, larger, decorative pot. The four-inch container was anchored by a decorative glass mulch we added to just an inch below the rim of the decorative pot. We then filled this larger pot completely with water. The result was a pretty rain lily, a pretty pot, and a watering system that required very little maintenance.

Our biggest source of rain lilies came from a failed experiment at a wholesale nursery. I had been working with a tissue-culture lab, and one day the head propagator called and said they had about four hundred one-gallon pots of *Z. grandiflora* that they could not sell. That was a call I needed, and I bought them all, only to find out that the bulb was in reality *Z. labufarosea*— another pink rain lily.

White rain lilies (Zephyranthes candida) growing in a boggy area in Louisiana. CLW

Rebecca with an armful of Texas bluebells. I find that the wildflowers (not a bulb) can be just as pretty as the bulbs. WCW

This particular strain bloomed in late summer. I particularly like it for its ability to survive with little care and make a showy bloom every year in late summer. Those are exactly the kinds of plants needed on the Southern Bulb Company farm and the kinds of plants that customers want in their gardens. I still remember leaving for a talk one day and coming back three days later to an explosion of pink in the back corner of the field. The blooms were followed by seeds, and the population continues to grow to this day. *Zephyranthes labufarosea* was originally collected in high altitudes in Mexico, and many different strains have been introduced on the commercial market.

Back in College Station with Rebecca, it was *H. robustus* that had my attention, and so did this woman collecting seeds with me in Cynthia Mueller's front yard. I sent the seeds we collected to a greenhouse grower who was aiding me in propagation and put Rebecca on her way back to Huntsville for class the next morning. I went back to the cabin; I was alone again and wondered about the future.

CHAPTER 5

Spider Lilies Say "Louisiana," and Oxbloods Say "Texas"

In many ways I am glad I did not meet Rebecca during the earlier years of my business. While I have no doubt she could have handled those building years, I am glad she was spared the years of 2006 to 2009. During those three years, I learned some of my hardest lessons in life: lessons about business, bulbs, and people. And I learned about markets and that markets could make very steep declines.

As time passed, I began to see in the business what I see in bulbs. During the life cycles of both, there are periods of dor-

A single red spider lily (Lycoris radiata). CLW

mancy, bloom times, and occasions of intense activity. Even though bulbs experience times of hibernation, I find that at least one kind of bulb will be blooming each month. In Texas, our intense heat and sometimes dry summers can cause plants to slow down, grass to brown, and people to stay inside, so we don't always realize the changes happening in our gardens. Summers in particular can be long and exhausting, but just when we think there is no hope (another similarity with business!), a late-summer rain arrives and awakens the surprise lilies.

    These bulbs go by many names, including "spider lilies," "British soldiers," and "hurricane lilies." Appropriately named "surprise lilies" because gardeners forget the bulbs even exist, and then flowers suddenly appear, devoid of foliage—just blooms popping one to two feet above the ground; "spider lilies" because their long, slender petals unfurl, looking like the legs of a spider; "British soldiers" because they appear to stand at rigid attention

These bulbs are sometimes called "British soldiers" because they can be seen marching across meadows in their red coats. CLW

in their red coats and in large masses preparing for action; and "hurricane lilies" because they bloom in hurricane season, sometimes after devastating rain and wind.

In the Gulf states, hurricanes are a natural part of almost every fall season, and while not always catastrophic, hurricane season brings uneasiness to many residents of the southern United States. When Hurricane Ike hit the Houston area in September 2008, Kathy Huber reported this situation in the *Houston Chronicle* on September 19:

> The hurricane lilies are blooming.
> My front garden looked like tossed salad on Monday morning thanks to Hurricane Ike, but the lilies, tough little bulbs, greeted me through the debris.
> The deep-crimson flowers of *Rhodophiala bifida* pop up each September. They're joined by their cousins, the spidery *Lycoris*. Both arrive at the height of the storm season and share the common name hurricane lilies.

This week I took their return as a sign I would have a garden again—and I have been in great need of reassurance.

We often see these bulbs with a mixture of sadness, joy, and hope, silently collecting memories to share with generations to come through their blooms, fragrance, and foliage. Perhaps the *Lycoris* can be considered one of the best storytellers of all, because of their dramatic blooms and timing. Even more exciting, they don't come just in red.

The lilac-pink *L. squamigera* bloom in late July and are often called "naked ladies." "Naked" refers to the fact that the bloom appears on a stalk without any foliage (the foliage comes and goes quietly and quickly in the spring); and "ladies," to the graceful, ladylike pink, trumpet-shaped blooms on the one- to two-foot stalks. I usually avoid the name "naked ladies" altogether, but at least one member of every garden club would feel remiss if she didn't raise her hand to start the crowd howling and me blushing by looking at me and saying, "We call them 'naked ladies.'"

Naked ladies bloom sometime after the Fourth of July and do well from the upper South all the way into areas such as Ohio. Some *Lycoris*, such as *L. aurea*, which has golden spidery blooms,

Bill and I sharing a good laugh about "naked ladies." CLW

grow near the coast. Other fun-colored, more rare varieties include the peppermint-striped *L. incarnata* and the more cold-tolerant yellow form, *L. caldwellii*.

My real education in *Lycoris* came after visiting further with Greg Grant. Greg has farmed bulbs on five or six little plots near his grandmother's home since he was a kid. Most of the stock came from his grandmother's yard, and year after year he diligently divides the bulbs. There were sizable rows of all sorts of heirloom bulbs. About two years into my business, Greg decided to stop farming and offered his stock for sale. Seeming like a perfect fit, and having relied on Greg for support from day one, I seized the opportunity to buy the bulbs and continue farming them on our land.

The Southern Bulb Company made a large jump at this point, and our inventory increased as well as our visibility. Through this purchase, Greg introduced me to some jewels of *Lycoris* (only the red *L. radiata* is normally referred to as the spider lily)— *L. caldwellii* and *L. incarnata*. I still remember digging the *Lycoris* plot. Greg's uncle came to check on me, advising me to enjoy myself that great spring day. After half the crop was dug, and after Greg's mom had cooked me a fabulous meal, I felt the effects of my 4:20 A.M. departure earlier that morning setting in. I found a nice cedar tree with green grass below it and took a nap.

Fischer, by then my constant companion, curled up next to me, and I felt a bit like Rip Van Winkle. Fischer was growing larger and larger, and the famed "gray ghost" breed seemed more like the "way too excited gray wiggly worm" breed. He was always excited to see me, and his rear end would wiggle, causing his whole body to shake with joy. He ran with me, ate next to me, and slept on the floor next to my bed. Everywhere I went, Fischer went.

So there we lay, and I began to fall asleep. Before I dozed off completely, something crawled into my ear and bit me. I think it was either an ant or some kind of beetle. It woke me up, and I resumed digging. Back home on the farm, we transferred the rest of these late-summer-blooming *Lycoris* into newly formed rows.

These were great *Lycoris*, but there are many more varieties. Where can I start to describe the many different colors and forms of *Lycoris*? Along the Gulf Coast, the cold-sensitive *L. aurea* thrives—solid gold, spidery fall blooms that will set seeds. Farther north in the South grow pure white, or white with a tinge of pink, varieties; some of these selections are called *L. alba*, and others are called *L. albiflora*. The petite, pink-almost-violet blooms of *L. sprengeri* are worthy of mention, though seen less often in gardens.

There is some confusion about the names of different *Lycoris* species, but *Lycoris* have also been confused with *Amaryllis belladonna* and *Nerine sarnienses*, even though they definitely don't fit into either genus. When in doubt, grow two of these genera side by side to compare the differences, especially in their ability to last and adapt.

Nerine sarnienses are also sometimes called "naked ladies," and they are the official flower of the island of Jersey, one of the Channel Islands. Therefore, they are sometimes called the "Jersey lily" as well. *Amaryllis belladonna* is the official flower of the island of Guernsey and is often called the "Guernsey lily." Both names are occasionally misapplied to the *Lycoris* varieties that grow in the South.

We did all right with the transfer of *Lycoris* from Greg's farm, with a few helpful suggestions from Greg. I came back later in the year for some of the more special *Lycoris* that were growing in a separate plot the goats were getting into—Greg called and made sure I got there before all the bulbs were eaten up.

Lycoris squamigera are sometimes called "fairy lilies," and here they grace a southern landscape in northern Louisiana. CLW

As always, throughout my wanderings many gardeners have willingly shared their bulbs with me. The pinkish-white *Lycoris* came from Natchez, Mississippi, from the garden of Southern Garden History Society members Bob and Glenn Haltom—they took me to eat catfish along the banks of the Mississippi River the night I was there. Their beautiful home was situated on the original site of "The Goat Castle," a famous Natchez plantation that burned many years ago. Another *Lycoris* came from the Birmingham home of then *Southern Living* magazine editor John Floyd. From an area north of Mangham, Louisiana, Deloris Dye, who worked at the local John Deere dealership, graciously shared spider lilies with me.

On one of the prettiest *Lycoris* sites I have ever seen, numerous red spider lilies were dotted across one of the oldest cemeteries in the South, in Natchitoches, Louisiana—how the Revolutionary fighters would hate to know that British soldiers were still marching all over the place! The most common of our surprise lilies, red spider lilies have a color that ranges from what some call coral to bright red on top of long, slender stalks. The description of the flower arrangement is in the name *radiata*: the flowers face outward from the stem in a radial fashion. The flower petals are best described as narrow, wavy segments, and the blooms have long stamens that protrude beyond the flower.

Once the blooms fade, the foliage appears and looks almost like *Liriope* foliage for the rest of the winter. Many folks plant spider lilies in beds with their daylilies. When one plant goes into dormancy, the other begins to grow, offering color and/or attractive foliage all year long. The red spider lily is the last of the *Lycoris* to bloom, as the rest bloom early in the year, beginning near the end of summer.

The *Lycoris* that has seemed to naturalize all over the South is believed to be an old triploid form descended, in some accounts, from bulbs from the Carolinas. First sent to New Bern, North Carolina, by Captain William Roberts, an officer who accompanied Commodore William Perry in the opening of Japan in the early 1850s, that spider lily has spread across the entire South, adapting to the Gulf Coast region and up into northern parts of the South. It is important to note that New Orleans also lays claims to the bulb and has its fair share of stands of spider lilies.

The year the New Orleans Saints won the Super Bowl, I found myself speaking at an event at the Audubon Zoo hosted by the two garden clubs of New Orleans on the day of the Saints' victory parade. Arriving in the city the night before, I could feel the excitement. "Who Dat?" was written on every vehicle, the mantra unifying the city. My hosts took me to dinner that night at Clancy's, a famous seafood restaurant.

We parked along the narrow, almost European streets of the Garden District and approached the restaurant, intimate lighting shining through the windows. The entrance was on the corner, and the dining room and bar were small. No seat was left vacant. To my right just inside the entrance were stairs that led up to a second floor.

Everyone was formally dressed, and the ladies had on the most extravagant dresses, with feathers, furs, boas, and hats. Our table was easily identified by the four-foot-high contemporary piece of floral art in the center and the six other members of the garden club awaiting our arrival. To my right was Tempe, from the Little Garden Club of Memphis, who had heard me speak before, as more and more clubs across the country invited me to speak, and my life as the Bulb Hunter became busier and busier.

Halfway through our noisy meal (the excitement from the streets spilled into the restaurant) all of Clancy's was on its feet, applauding, with cameras flashing. Walking down the staircase (the one I passed on my way in) was one of the Super Bowl stars, Jeremy Shockey. I didn't know who he was at the time, but by the end of the meal my dinner companions had me well informed.

Dinner finished, we went home, and I gave my talk the next day. People in New Orleans never give you just a good time; they go the extra mile, and a member of the garden club invited me that evening to watch the Saints' parade from one of the private downtown clubs. I marched in New Orleans Mardi Gras parades as a cadet at Texas A&M before Hurricane Katrina hit in 2005, but I had never seen crowds this large or this excited.

It so happened that my hostess was also in charge of the invitations for the Rex Ball, and my unit at Texas A&M had marched in the Rex parade. The Rex Ball fascinated me because it was through its records from years past that we have limited knowledge of flower bulbs coming into New Orleans.

In Monroe, Louisiana, a wonderful garden club hosts me every time I am in town. One of the members, Georgian Potts, was researching the origins of the spider lilies found in the surrounding areas. One day a student approached her who had connections with the Rex Parade in New Orleans that allowed him access to the records kept by each passing king of Rex. These records go back to the days when ships traveled between Africa and New Orleans. Uncovered in these documents were ship manifestos documenting the objects that were carried with the ships. Listed among the possessions traveling across the ocean were flower bulbs.

This would then fit with further evidence showing that many of the *Lycoris* sites still blooming today are located on former slave quarters. One Southern folk artist, Clementine Hunter, portrayed much of the cultural plantation life in Creole Louisiana, and one of her pieces portrays her interpretation of the spider lilies found blooming in those areas. However, even with all this information, we know *L. radiata* only as native to China and Japan, and we know of no reports of *Lycoris* growing in West Africa. The "flower bulbs" mentioned in the manifestos could possibly refer to crinum bulbs.

We now collect, propagate, and sell spider lilies for several garden centers and at the online store at www.southernbulbs.com. In Houston, our dry bulbs are for sale at the Arbor Gate. If you recall, the Arbor Gate is the nursery where Rebecca's sister worked, who put Rebecca and me together for a blind date! At one of my talks there, a garden club member asked me to speak to her group at a nearby residential community.

About a year later, I drove to the community in a very nice part of Houston with large homes. The living room where I was to give the talk was big enough to hold about thirty women. My contact introduced me and then told a personal story about our bulbs. She began, "I bought some of his bulbs at the Arbor Gate and planted them."

I cringed on the inside; I knew where this story was going. "No bloom, right?" I thought to myself. The main problem with *Lycoris* bulbs is that customers expect a bloom the first year, and no matter how many times I informed them prior to purchase or afterward, they still wrote the line I hate to see or hear: "I am so disappointed." Spider lilies are known for not blooming the first year they are planted, and even established clumps do not bloom every year.

She went on. "And as I was going to my shed to grab the pot and throw it away, I looked through my window and there was the flower, perfectly framed. It was absolutely beautiful." I took a deep breath of relief. I wasn't sure if she was telling me that the bulb was planted in her shed or if she was in her shed looking out, but I did gather one thing: her bulb had been planted after she bought it from me, and it had not done anything visible all summer long.

This was perfectly normal, but not to someone who understandably thought the bulb should start "growing" right away. The truth is that the bulb *was* growing. It was growing all summer long, just under the ground. I sold her bulbs that we had recently dug as they went dormant in late April, but the roots will continue to grow right through the summer. Most producers, in order to provide the bulb to larger markets, must dig the bulbs and let them dry out as they sort them. This can cause many *Lycoris* not to bloom their first year. The primary factors that affect bulb growth are temperature, moisture, and light. Temperature is often considered the most important. We notice temperature, as it marks the seasons, and we dress appropriately; bulbs notice and they grow appropriately, either beginning or ending growth with relative increases or decreases in temperature. In the South, moisture plays almost as important a role as temperature, as the rainy, wet winters more closely coincide with bulb growth.

The amount of light a plant receives is influenced by its location, and day length changes with the season. The foliage of a plant absorbs the light and turns it into food energy. Our "hysteranthous" fall bloomers are an interesting case and can be used to understand the effect of light on bulbs. Hysteranthous means that the flowers appear first and the foliage grows after the bloom, such as the red spider lily, oxblood lily, and autumn daffodil (*Sternbergia lutea*). These are interesting cases, because for most of the summer, when trees have their leaves, the bulbs themselves can be in the shade. When the bulbs bloom in the fall, the blooms can still be in the shade. When winter arrives, the trees lose their leaves and let the sun shine through. It is during the winter that the bulbs' foliage is up and growing. As a result, the bulbs receive all of the sunlight they need and convert the light into food for healthier, larger bulbs.

The fall-blooming *S. lutea* is one of the bulbs we see blooming in shady gardens in the fall. It is a favorite of rock garden enthusiasts, as it blooms closer to the ground and doesn't have the long stalk that other fall bulbs have. It is sometimes called the "autumn daffodil" or "fall crocus." Generally it can be seen in nice clumps along the Blackland Prairies of Texas, no doubt ensuring its survival because such environs remain drier for the bulb during the humid summer months. Many times, these bulbs may receive local names that include "surprise" in them, but the name "surprise lily" is most commonly used with the *Lycoris* varieties.

Most bulbs need more light to bloom than what is normally available in home gardens where there are many trees or where houses are built close together. Some bulbs in our barn will begin growing in the crates, grow very tall for a season, and even bloom using all of the stored energy carried over from the prior year. Leaves collect light energy from the sun through specialized structures and turn it into food energy through the process of photosynthesis. It is important to let the leaves on bulbs mature naturally, because during this time they continue to produce carbohydrates, which will aid the bulb in growth and flowering.

Nonetheless, many of our fall-blooming bulbs can be found blooming under the shade of trees such as crapemyrtles or pecans. Because the fall bloomers are hysteranthous, their foliage comes up in the winter; because the deciduous trees have lost their leaves, the bulbs have all the sun needed to stay healthy, propagate, and bloom. These characteristics, combined with our unique weather patterns, allow for other special fall-blooming bulbs. One of these bulbs tells an important story—it actually tells a couple of stories, about Texas and about the Southern Bulb Company.

Take a drive across Central Texas in the fall, and you will see a bulb blooming that is steeped in as much history as the black natural resource oil. In 1901, the Spindletop field near Beaumont began to gush oil. The black liquid from below the ground revolutionized the state and the lives of those who live here. It was a momentous occasion, but close to the same time, another black natural resource was making a stir under the ground. It was the little horticulture treasure known as the "oxblood lily," introduced into Texas about the same time as Spindletop but farther west in the Central Texas community of New Braunfels.

It was introduced by the German American plantsman Henry Oberwetter. He made a trip to Argentina and came back with a little black bulb with a long twisted neck most commonly known as the "oxblood lily" (*Rhodophiala bifida*). Some call it the "oxblood lily" because of its vermilion-red color, and other folks call it the "schoolhouse lily" because it blooms in the fall (around September) when school used to start. For those poor schoolkids, it meant it was time to go back to the classroom, but for all of us

Spindletop gushing oil. Photo by John Trust

in the South, it meant it was time for the blistering heat of summer to subside and lead into the cooler temperatures of fall.

Bill Welch has an oil painting hanging in his College Station home of a Victorian home in the little town of Calvert. The painting shows a scene from about the 1890s of a horse-drawn carriage, a large front yard, and a beautiful two-story house with a balcony flanking both sides of the house. That picture came to life one day, long before I knew it existed.

I was in Calvert with Natalie, with whom I had found the pink rain lilies (*Habranthus robustus*). We were of course talking about bulbs, and I was tapping her brain for some ideas of where I could find this oxblood lily. She knew of a home where they were growing in the lawn and made a few phone calls to inquire about permission to dig from the owner. The next thing I knew, I was walking toward a big house a few blocks away—the same house as in Bill's painting, only now it had been more than four times

A clump of oxbloods that has been in the ground for thirty years. It has never been dug, divided, or refrigerated, yet every year it continues to multiply and bloom. CLW

removed from the original owner. This bulb was in the yards of almost all of the old homes across Central Texas.

Because the oxblood lily could be considered the signature bulb of Texas, we decided to "take this bulb to Texas" and try to introduce it to as many people as we could. We called it the "Great Festival Tour of '06," and the experience tattooed a few sales and marketing lessons on my soul. The time had come for the Southern Bulb Company to grow, and more sales were what we needed. We hired Grant Cox for a second summer to help with this extensive sales push. Just think for a moment about the little red cabin at this point.

When I first moved into the cabin, it was just me and the rustic decor of deer heads, fishing poles, and dusty treasures. With Grant here that summer, the population of the 550-square-foot structure grew to four: Brad, Ben, Grant, and me. I really hadn't done too much to make the cabin better since I moved in. I removed the old carpet, but there were still only inches of hot water available for a bath and no hot shower. The cement floors still had old paint chipping away in numerous spots. Old doors with gaps at the bottom of about an inch still existed. One gap between the floor and bottom of the door was so wide that a snake slithered in one day. I could go on about the rugged nature of the place, but now visualize two people on bunk beds in one room, and the other two on fold-out cots in the middle of the main room.

One must also remember that Texas is hot during the summer, and humid. Your very essence begins to drain out of your pores along with the salt and sweat if you stay out too long. One puny little air-conditioning unit from 1970-something still hung in one of the cabin's windows, but the unit was not enough to keep us cool. We preferred to be on the porch looking out over the lake in the evenings. As a matter of fact, historically most Texans felt this way in the summer, and their desire to be outside led to social gatherings.

Many communities use the summertime to throw festivals and community events. Perhaps it is simply their creativity, or a warped sense of humor from the heat, but many of the communities have come up with some creative names for their festivals. In Aransas Pass, they have the Great Mosquito Festival. How appro-

Grant selling bulbs at one of the many Texas festivals. CLW

priate for our coastal city, since the early French explorer La Salle was "eaten" by mosquitoes on our Gulf Coast. In Luling, they have the Watermelon Thump, which of course has a parade concluded with the Watermelon Queen on her float; and in Lampasas they annually crown the Miss and Little Miss Spring Ho. These festivals, we thought, would be the perfect place to show and sell these truly Texas bulbs such as the oxblood lily!

"What a market we must be missing," we thought to ourselves. "Crowds of Texans gathered all across the state eager for something that was truly Texan. To target this market, we hired Grant. He traveled the state with our bulbs in hand, setting up a table or a booth as a vendor at all of these festivals. He was a hard worker, standing for over eight hours a day, under a ten- by ten-foot tarp, selling bulbs next to cotton candy vendors and local jewelry makers. When available, other members of the Southern Bulb Company team joined Grant in selling bulbs at these events.

Sometimes we were sandwiched between the guy with all of the flags and the face-painting booth. If we were really unlucky, our table sat next to a telecommunication booth. The sales reps continually walked out into the flow of traffic trying to get attendees to sign up for their plan. People would put on their thousand-yard stare and ignore the rep, along with the next three booths, just to keep from being hounded. We looked at each other in the booth and realized that we weren't selling any bulbs. We were hot, and sweaty, and tired of standing there.

There were some upsides to the festivals. The Fourth of July celebration in Granbury had a wonderful fireworks display, and the Schulenburg festival had a fun parade where many of the "top" festival floats from surrounding cities joined in the grand celebration. It also boasted the best Texas music and dancing. The Czech fest in West had a wonderful polka band and played plenty of waltzes, two-steps, and chicken dances. The Mecca of German festivals though is in New Braunfels in November—Wurstfest: A 10-Day Salute to Sausage.

Between festivals, I joined Grant if possible and sold bulbs on the road while exploring Texas further. Once we turned down a dirt road and found an old cross and creek behind a fence line marking the pool where Sam Houston was baptized—he said he felt bad for the fish downstream who had to deal with all of the filth washed off him! Back up the dirt road was the small church he attended.

Grant and I imposed upon some sofas and open homes as we traveled along. Jacob, who brought me Fischer, offered his parents' house along the coast when were down there for a festival. Grant and I squeezed in some crabbing and had quite the meal at Jacob's parents' house that evening. That was one of the many highlights among the disappointing festival sales.

We concluded our great festival tour in July, and Grant headed back to college. While we didn't sell nearly as many bulbs as we wanted to, Grant definitely picked up plenty of new and exciting stories, and our reputation across Texas grew. Through the process we learned more about what the best environment for selling bulbs might be. After this adventure, we decided that home and garden conventions were more our market.

Imagine walking through brush and debris behind an abandoned home only to stumble upon this patch of oxblood lilies! Photo Ben Arcuni

Conventions sounded like a much better alternative to festivals. They were all inside, which makes for a much more comfortable environment. Also, I was pretty sure there would not be a face painter in the booth next to me. Radio advertisements usually accompanied these events, and I knew that the gardening community would show up for them. I signed up for the Dallas Home and Garden Show.

The show turned out to be more home than garden, and most of the garden booths were related to lawn maintenance. We had interesting neighbors. Across from my booth was the amazing rubber broom infomercial continually playing, but it was a real person doing the same demonstration over and over. He would stomp hair and dirt into his sample carpet square, rake it to one side, pull the clean broom up to his eyes, and run his finger along the ridge proclaiming, "Cut's your time in haaaaalllllfffff." I had heard the talk about twenty times and saw more than one hundred rubber brooms marching out of the show in customers' hands. I hadn't sold a single flower bulb.

A poor kid spilled popcorn in front of the booth during a slow time, but no rubber broom seemed to be able to help solve the popcorn mess. I realized a lot about my market that weekend. I, of course, assumed that if everyone knew about the bulbs, everyone would gobble them up. It then dawned on me that most of America had forgotten about flower bulbs and that I was on a mission to remind them.

Across the board, however, we saw a softer reception to a bulb business than I had hoped. Interest in the life of the Bulb Hunter continued, but selling these bulbs to a larger market proved difficult. On top of this, the Southern Bulb Company now had a large overhead with inventory and employees, and the cash flow had not switched in the right direction. I could not go back to Joe and ask for more money. Southern Bulb was going to have to continue on its own.

Changes were about to happen, including some important changes in my personal life. Grant went back to school to finish his senior year and led an outfit in the Corps of Cadets at A&M, and I let other part-time employees go as well. Sadly, Brad and Ben were soon to be out of my life, too. Ben had an opportunity to pursue a higher degree and headed back to Texas A&M.

Bulbs tell stories of people and things that once were but are no more. Photo Ben Arcuni

Brad stayed on for another year but was pulled away by several factors. One important factor was that he and I both did not know if the company had the cash flow to support the two of us. Another factor was one much stronger than anything business related. A girl by the name of Katie from the University of Texas had entered his life. Katie Thorne and Brad fell in love.

Fischer was very curious about the bouquet of daffodils Brad held in his hand—daffodils recently picked from the bulb field. A long dog snout shoved its way into the flowers as I made Brad pose for a picture before he proposed to Katie. Brad wasn't going to ask the question on the farm, but he was going to that weekend on a trip to Seattle. He showed me the ring while in the office at the farm. Hearing the wheels of Katie's car approaching, Brad put the ring away. Brad gave her the bouquet of flowers. Making them pose together and taking another picture felt a little silly, but I knew they would be married soon. I also knew that Brad would be gone soon.

As the business matured, and we all grew older, it was harder and harder for friends to come help on the farm. The business was not growing as we had all hoped, and I could not afford to hire anybody new. Ben was gone, and Brad would eventually marry Katie and move to Dallas. Everyone was gone, and I felt the burden of a small business in a way I never thought possible.

CHAPTER 6

Our Most Valuable Bulbs: Roman Hyacinths, Byzantine Gladiolus, and Hardy Amaryllis

After the *New York Times* article, dozens of publishing companies contacted me. I even signed on with one agent to write a major national book about my adventures. Nothing happened because I did not write anything! I focused on growing the business, speaking to garden clubs as the Bulb Hunter, collecting bulbs, and turning the farm into a successful growing operation. As this opportunity to grow the business leveled off and all of my employees had left, it became quite apparent that I had neglected to write anything. Perhaps deep down I wasn't ready for a book; there was so much more that still needed to happen with my story. Furthermore, with the pressures of running the business resting solely on me, there was simply no time to write. Bill and John felt that writing a book was something that had to happen.

Bill Welch came into the area for a speaking engagement and brought to John's home some rain lilies and crinums. We all met the next morning, when they announced that various bulb collectors across the South who had generously shared their bulb collections and stories with me accused me of "nonwriting performance." Some of these "friends" approached Bill and asked him to convey their disappointment to me. In the beginning of 2005, I had begun blogging and people were hooked into following my life. When I stopped blogging, those people had become upset with me.

Clearly, Bill, John, and their supporters wanted me to start writing that very moment. John and Bill decided only they were allowed to determine when I had produced sufficient written material to be released from writing. Thinking to myself that this was silly and that I had other things needing to be done, I indicated that I didn't write well under such forced conditions, but they responded, "Take as long as you wish." Escape was hopeless, and after I had completed a couple of hours of serious writing, they reviewed the stories on bulb hunting and agreed to release me "on probation" until further notice.

Their serious condemnation awakened in me the sense that the business would always be here, needing my attention. It would not magically turn into a self-running machine operating without me, and if not careful, I would neglect the more important areas of my life. Nothing about starting or growing a business is quick— no silver bullet that will allow you to say, "Aha, I've made it! I can rest now." No, this bulb

Gladiolus byzantinus, *the Byzantine gladiolus, by British artist Ann Swan.*

business, I realized, was going to be a long, steady row to hoe, and one that might continue throughout my whole life.

Even though my schedule continued to grow busier and busier, with numerous garden clubs wanting to hear about bulbs, I found time to regularly sit down and write on my blog. I am now also keenly observant when my brother and Bill are together and make a point of being involved in all their plans and conversations!

This encounter with Bill and John was a turning point, and my writing improved and became more consistent. The blog was revamped by a third party, now that Brad was no longer around to help handle such projects. We set up another blog URL under the name www.bulbhunter.com and separated it from www.southernbulbs.com. I would use the blog to pass on horticulture information and tell stories about my triumphs and failures. It was through this blog that I was able to pass on my frustrations and successes with some of our most valuable bulbs.

Do you remember the old cartoons, where a rabbit mound is seen growing and moving in a line across a field until it hits Farmer Dale's line of carrots? All at once, the crisscross and swerving stops and the line becomes straight as it goes right down the middle of the row. From above the ground, we simply see carrots disappearing one after the other. That's what the gophers did to my Byzantine gladiolus.

After much collecting and growing, I had a pretty healthy stand of Byzantines growing in the field, and *Southern Living* magazine wanted to run a story on this prized bulb, among others. Not a problem. I'd just make sure I had the field looking good. Things had kind of gotten away from me, which can happen about the time April rolls around. Things in April always come upon me fast. The transition from January to April can be astounding. In January, even in our warmer climates, not too much is growing except for some hyacinths and *Narcissus*. More bulbs bloom in February and March, but at the end of March lawn grass also begins to grow. This green grass is the sign that life is about to burst forth, and in April we soon find ourselves behind mountainous weeds, battling a jungle of growth and high green grass. Such was the case with my plot of Byzantines.

Even though grass is competing with *Gladiolus byzantinus* in

A nice stand of Byzantines blooming in our field. CLW

mid-April, that does not mean you will miss the blooms. Two- to three-foot stalks packed with funnel-shaped, magenta flowers are not inconspicuous. It is one of the most powerful bulbs on the "heirloom bulb" list. The bloom is made of numerous individual flowers that are larger at the base of the flower stalk and smaller at the top. A common name for the bulb is "Jacob's ladder" because of the ladderlike appearance of the blooms. Perhaps most amazing of all is that these large flower stalks come from one small, acorn-sized bulb.

I stared out over my gladiolus field and wondered what could be done with the tall grass growing on each side of my Byzantines. Yes, there are farming techniques to help prevent this, but during the crucial time for weed control I was speaking to garden clubs, selling bulbs, and making other important connections. A fundamental issue kept nagging at me: trying to farm, market, broker, promote, collect, and botanize all at the same time was not working. Long-term solutions had been swirling in my head, but to solve the immediate problem, I rented a tiller and began churning dirt on the sides of the rows and weeding what was left by hand. The garden editor and photographer for *Southern Living* were coming the next day—this was backbreaking work, but it was my last option.

Before I left the field that night to prepare for the next day's visit, the foliage of one gladiolus caught my eye. I could almost hear the gopher calling, "Timber!" It was slightly yellow at the tip, and I tugged on it. I now had in my hand a freely suspended stalk of foliage with no bulb attached. I grabbed the completely green stalk next to it that had a long flower attached—up it came. I went down the row, easily lifting flower stalks and foliage as my stomach sank. It was a helpless feeling. Voles, not gophers, had gone down the row and eaten my bulbs in a day's time.

The writer and photographer came out the next day, but a total disaster was averted. We were able to take pictures with the reserve stock I had planted in black bulb crates.

We decided earlier that filling the crates with soil and planting them in the ground was an effective way to keep track of bulbs and keep critters away. The bulbs growing in the crates had enough rich magenta blooms for a great photo, and *Southern*

Living captured their story, but my bulb supply was diminishing, not growing.

My mind began to work overtime on the best ways to save the bulbs and bring the bulbs to market, both for the upcoming article and for the future of the business. Many of the gladiolus corms I planted in thick plastic containers and planted those in the ground. I took my remaining Byzantine gladiolus and put them in the hands of a specialty grower but kept some to sell myself that fall.

It was with horror that I discovered a year later that the varmints' strong teeth had gnawed through the bottoms of my thick plastic bulb crates. I emptied the crates in the shade of a tree on a hot July afternoon to harvest the bulbs but found none of the precious corms. After I inspected the crates, it was obvious what had happened. One perfect circle had been chewed through the bottom of the crate for the critter to enter, and another perfect hole had been chewed through the bottom to exit. That fall, I had to make apologies to customers who thought they had bought bulbs from me but that I no longer had.

Questions about gophers, moles, voles, and squirrels are a constant at every garden talk (right after I am informed about "naked ladies" of course). There are various ways the home gardener can control the problem, and I simply list them: plant the bulbs with chicken wire under and over them; plant with rocks; sprinkle coyote urine over the top; or mix in cayenne pepper and sprinkle over the top. I have also been told a simple chicken feed supplement, such as Gran-I-Grit, can be added to soil. The granite used to aid in a chicken's digestion of food not only improves the drainage of soil but helps deter voles. After one such answer to the question of varmint control, someone raised her hand. She told the audience she had tried cayenne pepper once and it hadn't worked; in fact, the squirrel got some in its eye and began to scratch at it . . .

The story went on into the gruesome loss of an eye as the squirrel struggled and scratched for relief, and it wasn't pretty. I kept desperately trying to stop her or interrupt her, but she was determined to tell her story to the end. An awkward silence ensued as we all tried to recover from the dark recantation. I thanked the audience for their time and concluded. Now, when someone asks a question about varmints, I simply say either plant bulbs that voles won't eat or plant bulbs where there aren't voles.

After my personal crate-planting disaster, I collected the remaining Byzantine corms. What had been thousands was now hundreds. The remaining corms were sent with their baby cormels to someone specialized in growing only *Gladiolus*. It was sad to see them go, after having worked so hard for them, but this solved several issues: (1) the bulbs continued to be kept alive and propagated, (2) I was no longer tied to farming them and could focus better on marketing, and (3) this established a better production forecast so we were able to bring down the price from twelve dollars a bulb to something much more palatable in the new hurting economy.

Yes, twelve dollars a bulb, a small acorn-sized cormel, and we fetched the same price for a Roman hyacinth, another striking bulb that was sometimes tasty to gophers. This elusive, ancient bulb could usually be smelled before it was seen and comes in blue, white, or pink colors. On many bulb hunts, I have searched all day and not found any blooming. I most often found it blooming in January in well-drained, partly shaded areas. The rocky soils around Austin seemed to suit it well, but I also saw it north of Monroe, Louisiana; in Lafayette, Louisiana; in Williston, South Carolina; and at least one other location in each of the southern states. Perhaps if the blooms were the bright yellow of daffodils or the bright red of tulips, they would be easier to spot, but they are harder to find. We consider them rarer than most of the heirloom bulbs.

After a long search for the bulb in Texas, it was in Cameron that I finally spotted a nice long row of white hyacinths. They lined the front walk of an extremely simple landscape. A carport on the right housed an old car. The mailbox showed a Hispanic last name. I know Spanish, at least enough, and walking up, I rehearsed my vocabulary and knocked on the door. A stunningly beautiful eighty-year-old answered the door and greeted me in perfect English. "Can I help you?" she asked.

In my stupefied way, I muttered out my story. She was delighted, wanted to talk, and did so, with her oxygen tank in

Georgetown White
Roman Hyacinth

This stand isn't the one from Cameron, but you can see how we collected the bulbs by location. We soon had names from all over Texas on our tags. CLW

tow. I listened. Her name was Alice, and she had lived in Cameron her whole life. Her father served in World War I. I was used to hearing about World War II, but not about World War I—"the war to end all wars." Alice offered me a few hyacinths, and we replanted her remaining bulbs nicely down the walk, cleaning out her flower beds along the way. When I returned a year later, the bulbs had survived but she had passed away. It was one of those sad reminders of the brevity of life. The feeling of the shortness of life continued to grow stronger in me. The thought that these abandoned places were at one time someone's personal garden oddly enough saddened me. It was a story that the bulbs, including the hyacinths, told all over the South.

This story is told less by hyacinths than other bulbs. What I mean is that they are harder to find. These flowers are reportedly found in groupings, or "mats," of bulbs gracing southern lawns. I assumed from reading about these clumps that I should continue to drive deeper and deeper into the South to find the bulbs. First I drove through Louisiana with moderate success, then on through Mississippi, with about two bulbs spotted, through Alabama, and up into Georgia. Continuing north, I made it into South Carolina, and finally, after long hours on the road, I spotted the prize. It was in a small town, and patches were coming up in the lawns of a few adjacent homes.

My first stop was to knock on the door of the home with the largest population of these little treasures. The gentleman who answered listened to my story and gave me permission to dig. My shovel was foolishly still sitting on my farm in East Texas, because the Bulb Hunter hurriedly left under the assumption that if he really needed a shovel, there would be a Walmart, Home Depot, or something similar close by. Those stores were not close by this small town, and the only shop in town was a little dollar store. Inside was a handheld mini-shovel on sale for about $1.50. Minutes later, the $1.50 shovel plunged into the ground and was leveraged back in an attempt to pop up a clump of bulbs. The metal in the shovel snapped immediately. Crinum clumps had broken my shovels before, but never a clump of Roman hyacinths and never on my first pull.

Desperately, I dug around the car for something to use. The search revealed a thick, long screwdriver that would be poor for digging speed and efficiency but would not break. I began wedging it in and popping up all along the edge of the clump. Eventually my hand could slide under the clump and pull the mat of bulbs back like a piece of carpet. The final roots gave way as I turned the mass of bulbs and dirt upside down. The turn of the bulbs helped safely rip open the clump down the middle and exposed the beautiful mass of purple bulbs that were the goal of this whole endeavor. My heart began to beat faster. It was a good find but consisted of only about one hundred bulbs. I cleaned up my mess, left bulbs in the new hole, and filled in the dirt. Driving back to Texas, I considered the trip a success, but at the same it struck me that a different course of action was necessary to get more of these bulbs. I was going to turn to Bill for help in finding more.

Bill has done quite well with the Roman hyacinths at his Fragilee garden, and in particular, he has done well with the white form. Of all places, I wouldn't have expected his to be thriving in the area of his garden that receives a regular amount of water that drips out of a pipe. The pipe is connected to a shallow well, and the water is pulled upward by the power of the picturesque wind-mill on top. Water spills out into an area with large rocks and well-drained soil, which would probably account for the reason the white hyacinths have not rotted in their dormant summer season.

"Chris, I have another friend you might talk to about Roman hyacinths," Bill said over the phone. He went on to tell me about Virginia Sue Barr, who lived in a stately French-Creole struc-ture just a half hour north of his Mangham home and who was a close friend of Diane's. This house was well known for its center esplanade, which was full of naked ladies (*Lycoris squamigera*), which bloomed on their stalks with pink trumpet-shaped flowers circling the top of the stem. Over the years the owner had also accumulated large masses of the old blue Roman hyacinth.

Her home was quite a sight and typical of many of the nicer homes in the area, with nothing but cornfields for a backyard. The grove of pecan trees leading up to the house made an appro-priate impression. Of course, my eyes immediately spotted the numerous clumps of hyacinths popping out of every available space. I knocked on the door, Virginia Sue answered, and we talked. Then we walked to the front to dig, carrying the lid to a cardboard box to hold the bulbs.

"We'll be able to get you a good start," she said as she put her shovel in the ground and began to dig. I was delighted to have her healthy and revered treasures, which had been handed down for generations in her family.

The shovel popped up its first clump with a good amount of bulbs. "Now, you also need a start of the naked ladies," she said. We walked to the front, and she dug several clumps for me. "Come inside; I want to show you a picture."

In the front entryway was a picture of her daughter in her wed-ding dress in front of a mass of red spider lilies, and to this day I still see a white dress with a clear image of the background of red blooms. As the picture showed, she had two types of *Lycoris* bulbs planted there! Virginia Sue managed the large home and producing farm with the help of her son, daughter, and their spouses and children. She walked me around the rest of the garden and gave me a dozen of another bulb often known as *Scilla* (*Hyacinthoides hispanica*). I expressed my sincerest gratitude but did not simply say, "Thank you." In the garden world you are never to say "thank you" for a plant—it supposedly gives you bad gardening luck, and this is very important knowledge if you are to practice the art of gardening. This was a great start, but I still needed to continue to find more to build up a larger stock.

Following additional leads, I drove to a little town outside Austin. I was told that a long row of this bulb could be found along a fence line of one of the houses in the town, and that is all I needed to start my truck heading that way. I took another friend with me, a friend from the Texas A&M Corps of Cadets and the same one who went on crinum digs with me.

Chad and I found ourselves once again driving across Texas, chatting about whatever struck our minds until we hit our destination. At that point our eyes were peeled, watching the ground out the windows to see if we might find what we were looking for. Hyacinths weren't in bloom, so spotting just the green foliage was going to prove to be a challenge. This undertaking was not as impossible as one might think. Only five houses of any permanence existed in this town.

It was for precisely this reason that our frustration mounted when we couldn't find the bulbs, especially when we surveyed the one house that seemed appropriate for such a find. The house had the customary hoop-wire fencing of the 1920s around its front garden. The house itself was obviously just as old, with an old concrete sidewalk and edging around the fence. A bar ditch separated it from the main highway through town, leaving a small grassy strip on the other side of the ditch and in front of this fence, the gate, and a concrete walkway. We crossed the culvert and drove the grassy strip parallel with the main road right up in front of the house, as conspicuous as could be. Chad looked

› *The blue Roman hyacinths (*Hyacinthus orientalis*) can be stunning. Sometimes they are called "French" Roman hyacinths because they can be seen blooming along the ancient Roman roads of France. CLW*

down out his window along the edging of the fence and asked, "Is this it?"

To my excitement it was! No one answered the door after a minute or two of knocking. A quick look around the property revealed a smaller road in the back leading down a hill to some open-faced tin barns and shops where metalworking of some sort was in progress. No one could be found at the bottom of the hill, but there was another road leading back up the hill in a slightly different direction. I followed this road, and it took me to the back end of a one-story shop, with a front that had a large window facing what could be called Main Street. It was here that I finally found somebody.

The resident listened to my story and understood what I was after. She informed me that her son lived there and told me what kind of truck he drove so I could be on the lookout for it. While waiting for him, I drove around the area for a while. On coming back to town, I found his truck in the driveway and went up and knocked. He couldn't have been a nicer guy and seemed to enjoy the company.

This gentleman had recently divorced and now had the house all to himself. The place was a little disorderly, but that was from a party he had thrown the night before. He had a tap coming out of a wall and offered Chad and me a beer. We accepted a small glass and began to chat. This was not normal protocol before digging hyacinths, but I was okay with it. We said our good-byes, dug some bulbs, replanted the row, and planted some Chinese sacred lily bulbs (*N. tazetta orientalis*) by the front entrance. They are all still there to this day.

Building up the numbers of hyacinths has proven to be difficult for me. I made several attempts at tissue culture, which involves sending the bulbs to a laboratory that specializes in cutting off small pieces of the plants and placing those pieces into test tubes. Of course, only certain areas of the plant can be cut from and only certain media (a medium is a growing substance like a white clear gel with certain nutrients) can be used in the tubes. It took nearly five years, but we eventually had some successes with this method.

Traditional reproduction of bulbs occurs in different ways. The most common, naturally occurring method, is asexual propagation, in which a genetically identical bulb is formed at the basal plate of the mother bulb; these little bulblets are referred to as daughter bulbs. A good example of this is a clump of daffodils that all formed out of the base of a mother bulb. Other bulbs produce seeds from their flowering parts, but these can take four to six years to develop into a full bloom. Onions, which are bulbs, are very often produced by seed production as well. Still other bulbs produce little bulbs called bulbils at the base of the leaves on their stem or on some flower parts. Tiger lilies will do this, as mentioned earlier.

Artificial propagation, such as tissue culture, is used very often as well. Also, new plants can be propagated by cutting the bulb into different parts, each part containing the necessary information and energy to form a new plant. This can be called chipping or twin scaling, and it is done in commercial daffodil production in Holland. Artificial propagation is generally used for newer varieties that need to hit the market very fast and in large numbers. After time, and for most bulbs, natural reproduction is fast enough to meet most needs. The Roman hyacinths, however, were too slow and rare for natural methods. In the end, tissue culture was the best bet for us.

I treasure these bulbs for several reasons. The heirloom Roman hyacinth foliage stays nice and compact and isn't an eyesore once it is finished blooming. No other bulb compares in color, especially during January and February, and the spicy fragrance reminds me to look down at the beautiful six- to eight-inch blooms every winter.

I developed a true love for this old "French" Roman hyacinth (*Hyacinthus orientalis*) and with joy researched more into the history of the bulb. A native of the Mediterranean region dating back to 1545, it was used originally to breed most modern hyacinths. The old Roman hyacinth bulbs are smaller than modern hybrids, but the small bulbs can send up two to three flower stalks and clump up nicely in a garden. The result is that a single clump can stay in bloom for over a month, starting in mid-January and continuing until the end of February. The blooms are also not what you expect when comparing with the hyacinth grown by florists, as the blooms are looser and do not form tight, compact rosettes.

They are often described as dainty, but there are so many dainty blooms all coming up together that the resulting garden impact can be spectacular. Some of the blues are so rich they almost appear violet, and the whites and pinks share with us beautiful colors as well.

The "French" Roman hyacinth gained the names "French" and "Roman" from where and when the bulbs are found. Apparently, one can find them trekking along old Roman roads in France, hence the sometimes added qualifier "French" Roman hyacinth. It is just one more connection this "New World" shares with the Old. The French influence also pervades areas of our culture, and this is nowhere more prevalent than in Louisiana. I know of one person housed in the woods of Louisiana who is the embodiment of the French culture.

This person does not celebrate French Independence Day ("Those were my people that were executed," he exclaims). Robert Smith speaks French and keeps a residence in the South of France, but on this side of the pond, he is perhaps one of the premier French antiquities dealers in the nation, specializing in seventeenth-, eighteenth-, and nineteenth-century French antiques.

My introduction to Robert came by way of Bill when both of us happened to be in Lafayette, speaking at the Louisiana State Master Gardener convention. First, an appointment was necessary, because Robert could very well be in France. Second, a hand-drawn map faxed or mailed was necessary to find the place. You know you have found his Acadian house when the gravel road leads you past majestic live oaks with Spanish moss dripping from the branches and the crapemyrtles marking the corners of his parterre garden come into view. Soon you will see the cypress picket fence, masses of old garden roses, and an obelisk in the middle of the garden.

We parked next to an old barn. I peeked inside and saw old doors and windows, with authentic hardware, and working keys from eighteenth-century France. Large, period olive jars sat here and there, as well as cloches—bell-shaped glass objects that were put over vegetables and flowers to protect them from frost and wind. Robert is known for being one of the finest importers of the vase d'Anduze to America.

Robert met us at the steps to his showroom, if it can be called that, and we began to converse about our lives and each other. As we talked, an appreciation and excitement for the occupation of each surfaced. By the end of our visit, I was excited about eighteenth-century French antiques, and he was excited about heirloom flower bulbs. We took a tour of the garden, and he dug for me blue "French" Roman hyacinths that he had buried near the crapemyrtles.

In fact, many bulbs that do well in France do well in Texas, like the species *Narcissus* and a particular red tulip. And hyacinths are used in the perfume industry in France. *Narcissus poeticus* is reported to grow in pastures where cattle graze. In the north of France and in the Channel Islands, crinums abound. The Channel Islands, such as Jersey, are actually British Crown dependencies, but there are many French street names and many French-speaking residents. The Southern Bulb Company would come to

A vase d'Anduze with heirloom purple bearded iris growing on either side. Photo Robert Smith

have dealings with a company on the island, and it would be the Roman hyacinth that would take us there.

The hyacinth even had its own mania in a period of history. This mania was not as great as tulip mania and occurred one hundred years later, but the bulbs sold for exorbitantly high prices like their cousins the tulips so famously did years earlier. It is easy to understand why. We even had our own little hyacinth mania going; John loved the bulbs as well, and he thought there would be value in hiring a botanical artist to illustrate them, along with our other bulbs.

"What we need are botanical illustrations," John emphasized on the phone. "Think about it. We're preserving these bulbs through collecting them and propagating them, but how did they preserve things in the old days?"

"How?" I answered, distracted as I finished typing an e-mail.

"Hey! Are you listening to me?" he asked. We were both quickly falling into the habit of e-mailing and talking at the same time. I responded that I was listening and stopped typing my e-mail.

"Through art. It will immortalize the bulbs . . . *and* . . . think about it. You're not telling me people wouldn't want one of these prints hanging in their study or in the kitchen or in their tea room. They can enjoy the bulb blooming 365 days a year." This was how many of John's and my conversations went. One of us would call and passionately explain our idea. After rebuttals and discussing, we decided to move ahead or not move ahead. The techniques of quantifying our ideas, putting them into time lines, and making sure they were profitable before we began came later in our lives after some rough roads.

"Well, I agree, but I'm swamped. I can hardly breathe, and the business doesn't have the money to go out and do this. So, if you want to do it, I think it's a good idea, but you're gonna have to take the lead."

So that is what he did. After doing research to find the best botanical artist in the world, he settled on a woman from London by the name of Ann Swan. Ann used colored pencils for her drawings and won gold medals from the Royal Horticulture Society. It was a long shot, but he saw her work and that was whom he wanted. John called her gallery in London, located in the nineteenth-century coach house in which she lived. Ann was vacationing in France, but her assistant gave John her cell number.

"Hello?" she inquired from the beaches of France. "You're who?" The "who" coming in an octave lower, signifying the English way of asking a question. "You're from Texas?" she inquisitively continued and then definitively concluded, "Well, I'm sorry but I don't draw *bulls*, just plants and other stationary horticulture items."

"Bulbs, like tulips, daffodils, and hyacinths." John explained as clearly as he could with the bad international connection. They were finally able to begin conversing, but the prospect did not look good. Ann had a five-year waiting list, and it was obvious she did only what she wanted to do and when she wanted to do it. Sensing the balance falling out of his favor, John concluded, "I would like to sit down with you and talk to you some more."

It is easy to get bulls and bulbs mixed up. This particular landowner said I could have all the bulbs I wanted. They didn't tell me they had bulls as well. CLW

Surprised that someone would fly to London from Texas, she agreed to meet. His decision to meet her in London sealed the deal, because he came back with a victory. Ann agreed to draw our bulbs, put her five-year waiting list on hold, and fly out to Texas. Ann had a great time in Texas but was shocked by the cabin. How could a story with so much press and nice pictures come out of such a place? We decided to show her how the cabin in the country afforded us to live.

We took her fishing and let her shoot a .30-30 rifle, "the gun that won the West." She posted the picture of herself firing the gun at her studio and informed her students, "This is what happens to you if you don't sharpen your pencils!" Later in the cabin, after Ann safely returned to the bed and breakfast over the hill, I pondered my situation. I suppose I was also shocked a little. We just had a delightful visit with one of the world's best botanical artists at the little red cabin on the lake.

Ann is well known for her illustration Bulbs in a Line *for the Southern Bulb Company. She is best known for being a world-class botanical illustrator.*

Oxblood Lily
(*Rhodophiala bifida*)

Twin Sisters
(*Narcissus x medioluteus*)

Byzantine Gladioli
(*Gladiolus byzantinus*)

Roman Hyacinth
(*Hyacinthus orientalis*)

Red Spider Lily
(*Lycoris radiata*)

Grand Primo
(*Narcissus tazetta 'Grand Primo'*)

While Ann was in Texas, we talked with her about the next extension of the idea of preserving the history and culture of these heirlooms, and it led us to the island of Jersey, with a company called Jersey Pottery, a third-generation pottery company. When much of the world pottery industry suffered, Jersey Pottery continued to grow.

Ann's work was incorporated onto the pottery, and as a result, these historic flower bulbs from the backyards of southern gardens made a splash in international circles. Some of the bulbs could not be considered rare, as they dotted roadsides and old homes from Texas up to Virginia, but all could be considered forgotten, at least in some respects. Ironically, many of the same bulbs that Ann illustrated are found growing in England. Her initial sketches included the Roman hyacinth, Byzantine gladiolus, red spider lily, *Sternbergia*, and her classic piece, *Bulbs in a Line*. She continues to work on sketching more bulbs.

One bulb in the works has a noted English connection and is commonly called the "hardy amaryllis" (*Hippeastrum × johnsonii*). This heirloom variety was first hybridized in England by a British watchmaker named Johnson in 1790 and is believed to be the first hybrid amaryllis on the market. All amaryllis will do well for us on the Gulf Coast, but this particular selection will survive into colder areas of the South such as Dallas, Nashville, and northern Georgia. Due to the widespread geographic availability, *H. × johnsonii* has various common names, such as "hardy amaryllis," "Johnson's amaryllis," and "St. Joseph's lily."

It is also one of our most valuable bulbs at the Southern Bulb Company, and some would say it is still the best hybrid amaryllis. The hardy amaryllis does not take long to multiply, and soon you will have a massive clump of the blooming bulbs.

Johnson's amaryllis blooms in mid-April. Modern hybrids have broader petals with a more flat and open face instead of the old-fashioned narrower petals and trumpet shape. St. Joseph's lilies are not modern in many senses of the word. Many varieties of amaryllis do not have a fragrance, but this hardy amaryllis has a fragrance if you take the time to kneel down and smell it. Newer selections are easy to propagate quickly, but the bulbs of *H. × johnsonii* are smaller and take more time to grow in numbers. Most amaryllis on the market are forced to bloom for Christmas, but this hybrid blooms in spring in the garden and sometimes again in the fall. It is, to date, the best garden amaryllis for southern gardens.

In the end, it was easier for me to send this bulb off to be propagated by a grower that specializes in growing only amaryllis. With

A nice large clump of hardy amaryllis we grew in a pot. CLW

a little irrigation, we could probably grow this bulb ourselves at the Southern Bulb Company, but keeping up with crops is a hard task when you are a one-man show. This technique of finding other growers is something I began to apply to more and more bulbs. It allowed me to focus on generating revenue to keep the business growing. In other words, I became a full-time salesman.

My life transitioned from a bulb hunter, to a bulb farmer, to a bulb salesman. In my passion for the bulbs, I assumed the only reason that the company was not growing faster was that more people simply didn't know about the bulbs. We had limited success selling the bulbs in retail garden centers, but I was convinced that customers all across the nation would buy the bulbs if they just knew about them. It was at this point that I decided that these heirloom bulbs simply needed to be in a big chain store, like Lowe's.

My brief attempt at big business. I was just sure the whole world would gobble up the bulbs. Apparently it takes a little more education and marketing than I had the capability of at the time. CLW

Navigating my way to a contact at Lowe's and convincing the buyer to purchase heirloom flower bulbs would take another book. Let me simply say that after years of effort and a sleep overnight in the Atlanta airport, our bulbs appeared in Lowe's across the South. Many people are quick to jump upon the horrors of large corporate box stores; I will say that the whole experience was positive. It required a lot of work, and I had to learn to communicate with brevity like I did not know was possible. My techniques and the process are not the point though.

What is the point of even mentioning the story of Lowe's carrying my bulbs, you may ask? The bulbs did not sell well. The buyer gave them a prominent spot in the stores and put them in good stores. When I received the e-mail from the buyer canceling the program, I realized another thing about my market. It took educating the consumers, and the consumer market interested in heirloom bulbs was not as large as I had hoped. The Lowe's experience matured my little business, and I became wiser through it all as well.

CHAPTER 7

Irises, Rock Garden, and Fischer

Throughout all of these experiences, my beloved Fischer accompanied me everywhere, including many of my bulb-hunting drives. As a puppy, he weaseled his way into my lap as I drove, something I broke him of later when he grew too big for such antics. On the farm, Fischer was right at home and curled up in a little bed made for him in the corner of the office.

With garden talk requests coming in, I ran into some dog-owner

Fischer smelling a crinum. CLW

Fischer even became part of the marketing material. This photo was used for much of our marketing for a while. CLW

challenges. Often I left him with friends in Dallas or Houston. One time I left him with a friend who had a Great Dane. Her Great Dane was so bothered by my dog that he scratched a hole through the drywall in the apartment to get out of the room he was in. Soon, Fischer's accommodations became part of the requirement for me to come speak, and I would walk upstairs to the guest rooms in some of the nicest homes in Texas with a way-too-excited Weimaraner in tow.

Some talks were just a little too far, such as those that included an airplane flight. Thus, I had to leave Fischer with friends when the Potomac Chapter of the Rock Garden Society invited me to the US National Arboretum to be their guest lecturer. At the time there was a national orchid exhibition in one of the museums on the Mall by the Capitol, and the orchids were all being grown at the greenhouses next to my presentation location. Never have I seen so many beautiful blooming orchids.

My host on the trip to the US National Arboretum was a nice couple from Alexandria, Virginia, both members of the Rock Garden Society, and he was a curator for the Smithsonian. On the afternoon of my arrival, I was delighted to have an invitation to visit the botanical library at the Smithsonian. An afternoon is not nearly long enough, but I quickly poured over documents, books, and more, researching one particular bulb that we had found growing across Central Texas. It was a red-flowering bulb that was soon to be a major part of my life at the Southern Bulb Company.

I think it is important to write a bit about Washington, D.C., and Virginia. Even though I have focused on areas of the Deep South and Texas, Virginia falls into the category of southern as well. Many more daffodils and tulips can be grown in Virginia than in a place deep in Mississippi, but Virginia is also home to many of the heirlooms previously mentioned in this book.

Washington, D.C., has always fascinated me, and during the summer before starting the Southern Bulb Company, I interned there. While some of my cohorts in this Congressional Internship Program ended up on Capitol Hill, I found myself at the base of the Hill, next to the Washington Monument, working at the Department of Agriculture. It was a very fitting position for me.

I was assigned to the administrator of the Agriculture Market-ing Service (AMS). His assistant had just been promoted, which left the office next door wide open with no one to occupy it. A well-timed entrance onto the scene meant that I had a desk, two sofas, TV, coat rack, and window that overlooked Independence Avenue. The administrator of the AMS is in charge of many things, such as buying excess fruit and vegetables and using it for school lunch programs or monitoring country of origin labeling (COOL) procedures—a law just passed when I arrived. That sum-mer, we took several trips flying across the country explaining and preparing for the complicated implementation of the COOL legislation.

As I described my internship experience with an individual from a major US bulb company—there had been a connection between my former job and one of the company's key employ-ees—she paused and looked at me with stupefaction and said, "Wow, you've really done it all."

I was kind of proud, as we all so vainly and quickly become with the slightest compliment. She continued, "But you're not there yet. You still have more you have to do." She was referring to my need to make the Southern Bulb Company a success. I was slightly deflated and reminded of my lowly position, but I knew she was right.

Thanks to the kind invitation of the Rock Garden Society, I returned to D.C. to talk about heirloom bulbs. My hosts allowed me to stay in their wonderful Victorian home in Alexandria, not far from a metro stop that allowed me to see the city and old friends at the Department of Agriculture. However, my time to visit with friends was short, as I was there to see gardens and give a presentation. The gardens around this particular Victorian home were splendid and had a wonderful collection of bulbs. They even had crinums and a very nice selection of irises.

Up to this point, I have not mentioned irises. I have hardly mentioned daffodils or tulips either, but they are coming soon. Irises are important for many reasons, but perhaps the great-est reason is that they are very tough and live in a wide range of areas. This garden in Virginia was filled with irises, and irises also fill just about every garden along the Eastern Seaboard. There are several different types of irises. Technically, irises are rhizomes; however, some irises, such as Dutch iris (*Iris xiphium*), actually form bulbs.

A well-known group of irises is the Louisiana irises, which can grow to be five feet tall. I have some of these Louisiana irises scattered around the pond at the farm. One is a nice, pale yellow, water-loving iris that Bill gave us from his College Station home. To fully go into all the different selections of Louisiana irises would disrupt this story too much, and it is sufficient to know that they love water, they are very beautiful, they are very quick to multiply, and you will not be disappointed by the blooms!

Most Louisiana irises are not actually native but are bred from natives. The most notable variety is *I. fulva*, which, oddly enough, blooms red or rusty-red and grows well in ditches and

The Southern Bulb Company logo with a fleur-de-lis. CLW

bogs. *Iris fulva* also seems to have adapted well to garden life. Like most natives though, it is very often passed up for more showy modern hybrids in cultivated gardens. Another important iris in Louisiana is one that is often mistakenly called a native.

The flower symbol of the New Orleans Saints is the fleur-de-lis, which literally means "lily flower" and has been the symbol for royalty, slaves, football teams, French-speaking people, and British nobility. We incorporated fleur-de-lis into the logo of our company, but at first the company didn't even offer any irises. This matter was called to my attention after presenting to *Southern Living* magazine staff during a planning meeting. The witty and well-known editor Steve Bender, "The Grumpy Gardener," asked:

"So, do you carry any irises?"

"No, actually, not right now," I responded.

"You do realize that the fleur-de-lis is an iris, don't you?"

I was embarrassed because I did not know, but it wasn't my first humbling experience. Technically, I could have gotten away with saying that several different plants have been considered the fleur-de-lis, but it is the general understanding, especially in the South, that the symbol is fashioned after *I. pseudacorus*.

This plant is often confused with native Louisiana irises, but it is not native. The common name is simply "pseudacorus," and it is a yellow-blooming iris that loves water. It can often be found aggressively naturalized in ponds and lakes along shallow banks. It is so aggressive that I fear to plant it on the farm. Some irony exists in the lack of irises on a farm that uses an iris for its logo, but I have seen it take over the entire shoreline of ponds. While on the surface, this sounds nice, it would not be long before it would drift down to the larger lake and cover the banks there as well. In the end, I would have nothing but shorelines of the irises.

A safer, more desirable, and less aggressive dryland iris that I often see is the cemetery white iris (*I. albicans*). It enjoys the drier conditions of cemeteries, elevated spots, or regions with less rainfall. Cemeteries are generally not irrigated and are located in elevated areas. This iris is not native to the United States, and I had an opportunity to discuss this with a fellow bulb enthusiast from Holland.

*The cemetery white iris (*Iris albicans) *blends nicely with an old selection of verbena. CLW*

My friend, Eric Breed, has been an unseen support to me throughout the years of the company. He was my main contact with the bulb industry overseas and assisted the company in supplying bulbs needed from Dutch growers. As the Southern Bulb Company site grew, the demand for traditional Dutch bulbs from the company grew as well. Eric was my contact to make that happen.

He came to Texas to visit with me and stayed in the little red cabin. My Dutch friend was extremely tall, and the bed I had for him in the cabin was quite small. He did not complain, and we had a nice dinner, drinks, and fishing in the lake behind the cabin. That night we discussed many of the bulbs growing on the farm, and he shared photos of bulbs he had seen around the world. One of the bulbs I showed him was the white cemetery iris.

"Yes, I have seen this bulb as well in Lebanon and in Greece and on many of the islands in the Mediterranean there," he responded.

"Really?" I asked.

"Yes. Do you know what they call it over there?" he in turn asked me with a smile on his face.

"No, I don't," expecting something very cultural and clever.

He said, "The cemetery white iris," and we both laughed. Sometimes plant names are universal.

The cemetery white is found all over the South, and there is one particular patch that is quite old and found in the old family property belonging to a lovely lady named Lucille, who lives near Brenham, a city situated between Austin and Houston. You might be wondering why Lucille enters the story at this point. Several years after the death of Diane, Bill Welch began dating again. He and Lucille fell in love, and they were married. With that marriage, a new home and family were in a way introduced into my life as well. She named her home Twin Oaks, after the two very large oaks situated on the hill next to her house.

The only flower that Lucille had in her garden from her grandfather was the white cemetery iris. One day Augustine, the gardener, dug a small batch of them along with some *Liriope*. Lucille, rightly so, did not want them thrown out, so we planted them in a prominent place at Twin Oaks. Twin Oaks is in the country and is closer to the small town of Independence than Brenham. Her family has lived and gardened there for four generations.

I had not put great effort into finding or propagating the white iris, but under a great deal of duress and pressure from Bill, I eventually acquiesced. It was so tough that it was not hard to grow. And he was right—something I hate to admit at times. Customers responded with overwhelming excitement when the bulb first came out on our Web site. We could not keep enough in stock.

In more northern parts of the South, we see another beautiful iris called 'Caesar's Brother.' It is a Siberian iris that does quite well. The best clump I observed was in the gardens attached to Tryon Palace, where I was asked to speak and stay for a couple of days. Tryon Palace is in New Bern, North Carolina, and was built in the late 1700s as the first permanent capital of the colony of North Carolina. Some residents there report New Bern to be the place of introduction for the red spider lily, but as previously mentioned, New Orleans also has laid a claim to the introduction.

Another nice clump of Siberian irises I noticed was on a trip to California, where I was asked to speak to a garden club in Los Altos, just south of San Francisco. From there it is close enough to drive to the coast and see Carmel-by-the-Sea and the famous Pebble Beach Golf Links. The irises, though, were spectacular in the gardens at the base of the Golden Gate Bridge. Spring really is a better time to see the bridge and gardens, when the fog is not so bad.

On this same trip I visited Filoli Gardens, a historic site for the National Trust for Historic Preservation. The house and grounds were built by a family who owned a significant gold mine in the 1800s. They knew how to spend their money on well-designed gardens and architecture. The gardens and mansion have an eighteenth- to nineteenth-century British feel. I was pleasantly surprised to see many of our southern narcissus in their gardens, as well as numerous irises. Irises can be considered plants that transcend all people, all places, or all cultures.

There are so many irises on the market that the sheer number can be overwhelming. I would suggest joining an iris society or buying a book specifically on irises to learn more. My best knowledge of irises continues to come from my personal experiences with them and the many places I have hunted for them.

One phone call and subsequent hunt led me to a peach-colored bearded iris that we are attempting to identify and propagate. The Clayton Library in Houston, a center for genealogical research, called and informed us that the center beds were going to be torn up for an irrigation installation. They asked if we would like their historic irises growing in those beds. Of course, I was down there to dig them up. This peach-colored bearded iris has performed with mediocre results on the farm. When they called a few years later asking how the propagation was going and asking if they could have some back, I was embarrassed to be able to send only a small box with the healthy tubers. I have done well with some irises and not so well with others.

For the longest time I kept about two dozen white irises in a crate by the cold frame, just down from the pole barn we had constructed on the farm. It was at this corner of the cold frame that I had the distinct pleasure of running into a copperhead snake. There seemed to be numerous copperheads at first on the farm.

One dark night, Fischer and I went for a walk on the farm, only to have him run ahead and check out something in the dim moonlight. The silly dog stuck his nose down to smell a copperhead, which, of course, bit him. Fischer always lacked common sense, and I feared that he would do something really stupid someday.

Fisch and I went outside to say good-bye to garden club group who had visited the farm. From my tractor, I told him to sit until they had driven off. He listened to me, and then I headed off in the tractor to the front field to pick up some pallets. Fisch ran ahead. As I turned the corner, I noticed him jumping along the bar ditch on the other side of the fence, the highway side. It was odd because Fischer doesn't go onto the highway when he's with me. It was an odd jump, too, and I thought, "Oh no!" but reassured myself that he was only jumping through the tall grass. It wasn't the tall grass; Fischer was trying desperately to get back to me.

He saw me as he rounded the corner to our entrance, and he jumped his odd jump all the harder. I watched him from the tractor, and he watched me as he took one more courageous jump toward me. He flailed in the air as if he had been shot. He landed in the puddle at the entrance and did not move. I jumped off the tractor and ran to him, saying "no, no, no" the whole way. I don't know why we always try to refuse to believe that some things can actually happen.

I lifted him out of the water and laid him next to me on our dirt road. He looked at me with frightened eyes, and I called his name louder. Then I knew, with his breathing and eyes that it was almost over. I just told him he was a good boy, over and over, and called his name. He tried so much harder to breathe. The final moment I yelled for him and sobbed. Muddy tears hit his face. His eyes dilated; his breathing slowed, then stopped.

I might not have ever moved. I heard footsteps and saw shoes next to me. They were the neighbors'. The kids across the way were staring. I had to control myself. I buried him in the woods, close to where we buried Red Dog, another dog I'd lost early in my life in the cabin.

To my friends who bought him for me, thank you. He meant so much to me, and I realize that all too deeply now. I miss him and can't shake the memory of his first day in the office, his scared face at the Christmas party when you gave him to me, the walks through the woods, the swimming out to the fishing boat so not to be left onshore, the night he was bitten by a copperhead and slept next to me on the floor, or every night that he barked at the coyotes, the way his whole body shook like a big gray worm whenever anybody showed up, or the wake-up nose when I slept too long, which included that morning.

Fischer was a gift for a short amount of time. My mom, who died of cancer during this same time period, was a precious gift that only now do I fully appreciate. Brad and Ben and the others were gifts to me as well, and they were no longer with me on the farm. The Southern Bulb Company had turned into something that was hard for me to manage and was not nearly as successful as I had hoped it would be. I was fairly well settled with all of these losses and transitions, but the sadness began to overwhelm me.

It is a funny thing that I speak so much of renewal of life with the bulbs. I see houses and land that have long been forgotten

One of my final photos of Fischer with John, both next to a milk and wine crinum. CLW

Fischer and the Bulb Hunter on the lake enjoying the fishing. Photo Ralph Anderson, Southern Living

but that were once homes filled with joy, excitement, and love. Should I find it strange or unique that I, too, should have to travel through sad times? My faith and hope did not waver during these times, but that does not mean that I did not experience extreme sadness. The sadness did not last forever. Sadness is a season that does transition into spring.

A friend and coauthor, Cherie, helped me realize that what is truly special about these heirloom bulbs is that even after everyone else has forgotten, the blooms still continue in their seasons to remind us. My horticulture adviser at Texas A&M put it another way by saying that bulbs are like old friends; you sometimes forget they are there, but then the blooms pop back into your life and bring you so much joy and happiness. For me, the season of sadness was coming to an end, and just as the long dark winter is ended by the daffodils blooming in the spring, my own daffodils began to bloom in my life.

CHAPTER 8

Seasons Filled with Daffodils

nots formed in my stomach as I took the highway exit for Huntsville, Texas. This was the biggest step of my life. It would be forever. Was I really ready? How much did I actually know about this girl? It had only been nine months since I first met her on a blind date.

I called the flower shop one more time to make sure the flowers were ready. They assured me they were, and I detoured off the

A field of daffodils in Georgia, along with Lent lilies. Lent lilies are also called buttercups (Narcissus pseudonarcissus). CLW

main road into town to stop by the flower shop, her former place of employment. The girls greeted me with big grins. More knots developed in my stomach. We conversed briefly, and I pulled my wallet out to pay.

"This flower arrangement is on us," the girl behind the counter told me. She continued, "We all think the world of Rebecca."

I thanked her and left with at least one hundred dollars' worth of flowers in my hands. It was almost dreamlike as the truck led me toward the Café Texan, the scene of our first blind date. Thirty minutes was a big cushion, but that is how much time I had to bring the flowers in, explain to the waitstaff what I was doing, and get the right table by the old window. Through the window was the courthouse, situated on the Huntsville downtown square. It was the same spot as our first date.

Inside the restaurant it was quite a lazy afternoon. One soldier was there in his uniform with a college girl having lunch, and the young waitress was cleaning a little bit here and there as she waited on tables. The time ticked on almost interminably until Rebecca appeared again, more beautiful than she was nine months earlier. Even though the flowers were out of sight, she was on to the gig from the start. We both hardly touched our lunches as I pretended through casual conversation that nothing was going on. Finally the time was right.

I got down on my knee and asked Rebecca Joy to marry me. She began to cry, and the girl from behind the counter brought out the flowers. The whole restaurant started clapping, and someone proclaimed that our first child had to be named Texan. I am not exaggerating. My life and her life were changing in front of our very eyes. Rebecca said yes. She said yes to so many things known and unknown. We were both filled with joy, and not long afterward the preparations began. Bright, yellow, cheery daffodils were blooming in my life after a long winter!

Our first wedding shower in College Station was a wonderful reunion of old friends, and my Dutch friend, Eric Breed, timed his trip to see me in order to make the party. Eric is an expert in narcissus and encouraged me to consider offering for sale some of the newer varieties of daffodils that reportedly do just as well in the South as many of our heirlooms. In the process of bulb hunting, one finds quickly that the most commonly found bulb is some type of narcissus. Daffodils belong to this group of plants. If they are the most commonly *seen* bulbs, then it is correct in this case to assume that the most commonly *dug* bulbs are narcissus. Bulb hunting and narcissus go hand in hand.

A few have performed well enough to be considered perennials in many warmer climates: 'Grand Primo,' 'Erlicheer,' 'Falconet,' 'Jetfire,' and 'Golden Dawn.' The larger-flowering daffodils include selections such as 'Ice Follies,' 'Fortune,' and 'Carlton.' Thousands more cultivars exist, but these are some of the most successful and popular. In the meantime, I focused on heirloom narcissus, traveling those abandoned roads, looking for the narcissus that had been growing for one hundred years and covered landscapes in carpets of gold.

As a guest at one garden club function, I found myself balancing my drink and plate of pound cake, strawberries, miniature sandwich, some chocolates, and cheese balls. Over the last several years, this food grouping has served as my fare more often than I think is healthy. A woman approached me as I ate my lunch. I could see the look in her eyes; she wanted an answer to something.

"I need to talk to you," she said matter-of-factly.

"Okay," I responded.

"I have a question," she continued.

"Okay," I answered, still balancing my uneaten food.

"I want some of those old-fashioned daffodils that you see on old homesites," she said.

"Okay," I answered.

"Now I don't want those little ones; I want those big ones that you see just covering some of these places."

"Okay," I answered, but this time I began a series of questions to narrow down just exactly what bulb she was referring to. It turned out the bulb was a Lent lily, and she continued:

"You're going to think I'm crazy, but I actually stopped by the side of the road to dig some up." On the inside I laughed.

She must not have heard about what I do for a living. However, this brought up another point that I didn't go into at the time, because I wanted to finish my food—the point being that every piece of land is owned by somebody, and technically the roadsides as well.

A good way to test the validity of this statement is to take one step over the property line of some place that looks abandoned. In the best-case scenario, someone will drive by suspiciously and stop to chat with you until the person is certain he knows exactly who you are and what you are up to. In the worst-case scenario, a farmer with a shotgun will appear seemingly out of nowhere. He will be less interested in asking you questions and more interested in you getting off his property.

Bill was about to cross a property line for pictures of a landscape formed of old plant materials when a truck pulled up. "I'm just taking some pictures," he quickly explained.

"Oh, that's fine then. I just wanted to make sure you weren't hopping the fence."

"No, just taking some pictures," responded Bill, glad the truck showed up before his camera and person decided to get a closer shot of the flowers.

And they meant it. CLW

Roadsides and the land in between roads are most often owned by the state; at least this is the case in Texas. Laws do change, but around the year 2005 the best explanation received from my numerous calls to the Texas Department of Transportation was that it was illegal to dig anything from the roadside, not just blue-bonnets (the state flower of Texas). However, if a piece of land was under contract from a construction company, the construction company had the right to grant me permission to dig from the land.

Along a highway expansion project in northeast Texas in 2006, I spotted a hill full of jonquils. The blooms of a jonquil are small, about the size of a dime, but there are three to five individual flowers on top of each stem. The flowers are pure gold, and their fragrance is pure sweetness.

Their scientific name is *Narcissus jonquilla*. They multiply through division or by seeds. When a bulb has the extra option of multiplying by seeds, large fields of them can be found. The sweet carpet of gold covering this hill could be considered normal but could not be called anything less than spectacular.

After researching the situation, I found the information for the contractor in charge of the construction on this stretch of road and called him on my cell phone from the patch of bulbs. He pulled up in his truck. I cringed as I saw him rolling over the flowers as he drove up to me. With the truck still running, he rolled down his window:

"Son, what is it you want?" he asked. I explained who I was and what I do. He concluded "Look, in two weeks, this hill is not going to be here, so you can have what you want."

That is all I needed to hear, and Brad and I began to dig. We were at the same site digging the next day when the contractor drove up. This time he steered clear of the jonquils! To my surprise, he turned the truck off, got out, and began to chat with us for a while. After some chitchat, kicking the dirt, I realized we were to the point of our conversation when he remarked, "Say, I talked to my wife about what you were doing. Do you think I could have some of these?"

"Absolutely!" I replied as I found a nice clump and threw them in the back of his truck.

The equipment was broken; otherwise, we would not have had the few days to dig the bulbs. CLW

"You know there is another house up the road that we're going to tear down soon. You might want to go look around there." I thanked him very much, and it was then that I realized something special had happened. We had won a convert to the joy of flowers.

Never again would this person view the world or see the beauty of the landscape in the same way. Many narcissus can have this effect on people, stunning them with vast displays of their golden-yellow colors. In reality, not too many selections of narcissus can be found in such large masses in the South.

It is easiest to start with three simple varieties when trying to sort them out: the jonquil, the Lent lily (*N. pseudonarcissus*), and the 'Campernelle.' The jonquil I have just described sets seeds as well as multiplies through the process of division. When setting seeds, it very often shares pollen with the Lent lily, the other seed-producing narcissus.

The Lent lily has a large trumpet-shaped bloom, the bloom that we identify with daffodils. Daffodils are in the genus *Narcissus*, but the common name "daffodil" is given only to those with this distinctive trumpet shape. The easiest way to identify the Lent lily is by examining the bloom. The cup of the bloom in the center is longer than the petals that surround it when they are both pulled forward together. Both the petals and the cup are bright lemon-yellow. The trumpet size varies because the seedlings from these flowers exhibit different genetic qualities from each parent.

When the flowers of the Lent lilies mix pollen with the flowers of the jonquils, the result is the third and most common narcissus, the 'Campernelle' (*Narcissus* × *odorus*), called "giant jonquils" by some people. As a result of both parents, the 'Campernelle' has the fragrance and bloom type of the smaller, dainty jonquil, but the bloom is larger and the foliage more flat, like that of the Lent lily. They seem to be slightly better adapted to different climatic locations than either parent, like a mule seems to be a better work animal than a horse or donkey.

These pure yellow and gold narcissus are a relatively small sampling of the number of narcissus that exist in this world. The paperwhite, another one that is well known, is pure white and is normally seen blooming around Christmas.

In Dallas there is a nice home with a garage apartment in the back. To reach this apartment, I often enter through a heavy iron gate that leads into the backyard. Although very heavy, it turns almost silently, and after entering, you must walk through the garden to reach the entrance of the apartment. The apartment is more like a small home than just a place over the garage. Sue and Phil, the owners, were gracious enough to give me a key so that I always had a place to stay while passing through Dallas.

The garden is splendid and often filled with paperwhites in December. Paperwhites are familiar to many people, even if they are not gardeners. About half of the general public like the smell, and the other half do not. Most do not realize how closely related to the daffodil paperwhites are.

All paperwhites are known as *N. tazetta papyraceous*. They are pure white and have numerous smaller flowers clustered together on top of one stem that give the appearance of a solid white globe.

The fragrance can be overpowering, but bulb-hunting adventures from December to early March will lead to the discovery of numerous other *N. tazetta* selections that are not as fragrant, and some of which are better for the garden. They cannot be classified as paperwhites though, because even though the petals are white, the center cup is very often yellow or orange.

Many of the selections of *N. tazetta* with a little yellow in the cup are better suited and stronger bloomers for garden-type settings. These are old bulbs and have picked up many names throughout their existence. Such bulbs include the Chinese sacred lily (*N. t. orientalis*); its double form, the 'Double Roman' (*N. t. romanus*); and the Italicus (*N. t. italicus*). Among these, later-blooming varieties are still favored due to resistance to freezes, including 'Grand Primo' (*N. tazetta* 'Grand Primo'), 'Grand Primo Citronaire,' and the double form of the 'Grand Primo' called 'Erlicheer.' These all have a sweeter, citruslike smell. There are many more, but these are some of the most well known. All of these varieties will come back year after year in warm climates.

Sue's garden in Dallas has many of these flowers, and it was in this garden enclosed by a fifteen-foot stone wall that I found myself one cool spring evening. Rebecca and I were there for a wedding shower hosted by Sue, and we were surrounded by some of our dearest friends from the North Texas area. The garden that had served as a respite from my weary bulb-hunting adventures for so many years served that evening as the perfect scene for our shower. This evening was a clear example of how my business life and personal life had merged into one. Perhaps they had never been separate.

I wish you could have smelled the jasmine in the air and felt the perfect breeze of the evening. The familiar trickling of the small fountain in the pond greeted us, as well as the fragrance and sight of beautiful flowers, shrubs, citrus, bulbs, and more. In this pleasant scene, the party lasted well into the evening. The crowd delayed their departure as long as possible, and Rebecca reluctantly left me to stay with her sister in Fort Worth. I decided to sit down on the bench and enjoy the evening before returning to the guesthouse.

On the farm that particular year, many of the narcissus bloomed late, but some years, selections such as the Chinese sacred lily will bloom much earlier. During Thanksgiving week one year, our Chinese sacred lilies and 'Double Roman' bulbs were making quite a show and releasing their delightful fragrance all across the farm. They had been forced by nature, so to speak. We had had a long, dry summer, followed in late August by steady moisture and cooler temperatures. Nature timed them perfectly to bloom for Thanksgiving. It was one of the most spectacular displays of blooms from the *N. tazetta* flowers that I had seen.

Chinese sacred lilies are wonderfully bright and cheery, and when Bill offered to let me thin some at Fragilee, I decided to do it. Unfortunately, the only time I could squeeze it into my busy schedule was late one night in January, so I had to use the headlights of the truck to see what I was doing. Even when in the country, digging bulbs by truck light draws some amount of attention, so I made sure I called the neighbors across the way

The fragrance of the Chinese sacred lilies (Narcissus tazetta orientalis) *is something I very much enjoy and associate with my earliest days of bulb hunting. CLW*

to let them know what I was doing. In the end though, with all of the chatter that goes on in this world, and with no small help from Bill, it has now become accepted as fact that I have stolen bulbs by truck light late in the evening!

In January, the Italicus, a more consistently timed bulb, blooms in cemeteries. I do not dig in cemeteries, but cemeteries are great indicators of what bulbs might be seen blooming in the surrounding areas. In a cemetery south of Dallas, I have seen strong stands of this bulb blooming year after year every January. Even though it is apparent that the foliage has often been mowed over and cut with weed whackers, the flowers still come up and bloom on solid stalks that do not fall over. A little lemon-yellow cup fills the center of nearly every star-shaped, creamy petal bloom. In some of the old literature, it is said that this bulb can be seen growing in the gardens along the banks of the French Riviera—an exciting thought that this bulb can also be seen growing in many gardens across the South.

'Grand Primo' bulbs have been coming up on this abandoned site in central Mississippi for at least fifty years. CLW

Perhaps the best narcissus selection for the garden in most areas of warmer climates is the 'Grand Primo,' and it is one I'd often searched for. Driving over an overpass on a highway just north of Houston, I spotted a row of white blooms. They were nicely contrasted against the gray of the cold late February afternoon, and the white puffballs of blooms were supported by the strong green clump of foliage that had emerged through the pile of leaves. The row was along a long cement drive leading up to a home, and judging by the condition of the landscape, the leaves from the previous fall had most likely been pushed to the side by the wind and not by any human.

These closer observations were made only after I had driven half a mile up the highway, exited, turned around, and driven the feeder road for another two miles. The blue truck's poor brakes, gears, and engine had been severely taxed by this point, but they were going to have to hold up. I could not afford a costly repair at this stage of my life or business. They all worked together to keep the truck moving as I continually drove bad roads.

Once I found the house, I saw that the white blooms were those of 'Grand Primo' bulbs. These blooms were characterized by their robust foliage, late February blooms, and numerous clusters of flowers with small white petals and lemon-yellow cups. The fragrance that hit me was also a dead giveaway—almost like that of a paperwhite but sweeter, like the Chinese sacred lily and 'Double Roman.' Among *N. tazetta* varieties, they were undoubtedly some of the very best. Even a small bulb would send up one, sometimes two, blooms, and the bulb itself would be sure to propagate more for the following season.

I drove past the house, looking for any clue that might tell me the nature of the person with whom I was about to engage. The yard was kept down not necessarily from mowing but from the cold winter temperatures. It was a large house that looked as if it had once been grand but was perhaps no longer occupied by the original owners. To the side, the cab of a big-rig truck was parked.

I deemed it was safe and discerned what I thought was the entrance most used. I walked up, looked for a doorbell, and finding none, I knocked on the door. There was no answer, and I began to walk away. The door cracked open, and a gentleman about forty years old poked his head out. His eyes almost squinted, and he wasn't shaven. He was, I believe, a truck driver, and my knock woke him from his sleep on one of his days off. I felt terrible for having woken him up.

At that point I really just wanted to leave. Leaving would have been even more awkward, so I told him my business. He said I could definitely have some of the bulbs along the drive, and that was about all he had to say. I think he wanted to get back to bed, so we said good-bye and I began digging from several of the clumps.

I knew these collected 'Grand Primo' bulbs were the best for repeat, increasing blooms for years to come in the home owner's garden, but for the time being, the major market for paperwhites will be the "forced" market. Forcing is a process of timing a bulb to bloom early or on a set schedule for some preconceived reason (i.e., paperwhites in time for Christmas). It usually involves drying a bulb down to a dormant state and then storing it at different temperatures for different amounts of time.

After this period is finished, bulbs are planted and will bloom at the desired time. For paperwhites, many can be planted and will bloom about twenty-eight days later. Most paperwhites that we force to bloom in pots are grown in Israel. They have names such as 'Ziva' and 'Bethlehem,' but there are new varieties that have stronger stems and sweeter smells, such as *N.* 'Inbal' and *N.* 'Ariel.'

Understanding the world market of flower bulbs became a major part of my life dealing in bulbs. Soon I realized that I could hand many of the bulbs off to other growers and have them grow them better than I could, as much of my time was spent on the road speaking to groups about bulbs or looking for bulbs. By gauging our Web site and audience response, I began to see which bulbs sold better, performed better for customers, or solicited a better response from the audience when I would flash large pictures on the screen. It was a big deal for me to realize these simple things about my market. It didn't matter what I liked; it mattered what the customer liked. With fine-tuning, the Web site revenue began to increase.

It was a turning point in the business. I had cut costs and focused on the market. The Southern Bulb Company began to make a profit. This turning point was one of the most exciting moments of my life, even though I cannot really peg it to one specific instance. The change was gradual, and it developed over time.

I had to find what customers liked, and our market responded. There remained one bulb that all customers would want that I still had not offered. Remember that my story began with the search for a red tulip that would perform well in southern soils? As I thought of the early years of the business, John's words almost haunted me: "You need a tulip. You need a tulip." We had looked for a tulip. We had searched for a tulip. We drove all over for Texas looking for a tulip. Could we really find a fabled red tulip that would come back in the state of Texas?

CHAPTER 9

The Cabin Becomes a Home

"**S**on, what's the matter? I haven't heard from you in a while," Joe said when I called him to give him an update on the company.

"We've been busy!"

"I picked up the *New York Times* piece. We're real proud of you! I showed it to everyone who didn't want to invest in you down here." I didn't respond to Joe's comment but continued to make observations about how lucky we were to have the piece written.

This tulip is on South Congress Avenue in Austin, but unfortunately it truly is larger than life. CLW

Joe, however, was in failing health. Two short weeks after he read the article, Joe died. He never met Rebecca. I was sad to lose him. When Joe died, another investor pulled out. I found myself not far from where I had started, but after surviving hard times, the business and my life came out on solid ground.

Rebecca would go on a few bulb hunts with me before and after our engagement. Bill continued to offer his assistance and remained adamant that some plants needed to be on the market, threatening to farm the bulbs himself if I didn't get around to it. One of those bulbs he was adamant about was *Crinum* 'Mrs. James Hendry.' He had a population at one of his old residences. The new owners had graciously given him access so that he could divide some of the bulbs. Bill invited me to come dig with him, and I figured this was something Rebecca should experience with me before we tied the knot.

"Oh, well," I thought to myself. "She had better experience a crinum dig now and see if it is something she can get used to." I hoped she wasn't thinking this was going to be anything like collecting rain lily seeds or white irises.

Thus, Rebecca Joy came with me on a very hot bulb-hunting adventure. Dirt and sweat abounded that day, and it was certainly not the cleanest dig I had ever been on. To my great joy, she was all smiles all day long, which is when I knew we were going to make it.

Some important bulb hunts occurred before Rebecca was in my life. I have not yet mentioned them, but at this point you might have guessed that I have saved my search for the red tulip to the very end. There were several tulip locations floating in Bill's memory, and one of the first ones he told me about was a site of a red tulip he had seen as a freshman student at Southwestern University, just north of Austin in Georgetown.

In Georgetown he recalled a prairie-style home just across from the campus, where he saw a six-foot-diameter patch of large-flowering, spectacular red tulips. That was in 1959. "Could the tulip really still be there?" I pondered as I first set out to find the tulip in some other areas of the state. A trip down to Austin was out of the question for the time being, and I focused on some closer locations.

This tulip becomes truly stunning as it opens up. CLW

At first I started in a small town south of Dallas. After a few small finds and tips from locals, I found myself arriving at the run-down remnants of a home, littered with the red-blooming tulip. My excitement began to surge. Walking up to the closest tulip, I bent down for a closer inspection. The outside was red, and on the inside the bloom matured into a deep red, with intense colors comparable to the berries of the Christmas holly. Poured into the bottom of the inside was a black pigment, dark enough to give the appearance of a pool of black ink sitting in the bottom. Lining the black was a pure, bright yellow, solid line that traced the top edge of this pool of black ink and separated it from the brilliant red to finish off the coloring.

A parked truck at the edge of the lot let me know that I was not alone, and I took it upon myself to approach the person standing by the truck looking at the house. I spoke to the stranger and asked about the bulbs. He replied in a gruff voice, "You're too late, son; I'm tearing this place down in a week."

I didn't care about the house, and I knew they could tear it down and the bulbs would be fine. What I didn't want to happen was for them to bulldoze the place and lay a foundation over the bulbs for a new house.

He kept talking as if he was the final authority, but I knew what to ask.

"Do you own this property?"

"Well, no, I don't."

"Could you tell me who does?"

After some more shuffling around the issue of him not wanting to relinquish control of the situation, he gave me the name of the owner. I called and the owner had a different tune. Thankfully, she was thrilled to relinquish the bulbs if I would just bring some by her home so she could plant them there. I began digging.

Immediately I noticed something different about this dig. The tulips weren't in clumps of fifty or one hundred bulbs, like so many of the other clumps of bulbs I have dug. They were all spread out individually, about three to five inches apart, and they

The tulip is phototrophic, opening and closing with the sun. This picture was taken in the morning. CLW

were each about six inches to a foot deep. The spreading nature is indicative of their underground stolons—they would send out stolons, almost like running roots, and produce another bulb not far away from the original. A stoloniferous bulb means one thing to a plant collector; every bulb was going to have to be individually dug in some of the hardest soils the state has to offer.

I stuck the shovel into the Blackland Prairie clay, and it went down only a couple of inches. I jumped on the shovel, and it went down a couple of inches more. A bead of sweat had already landed on the dry dirt. I popped the bulb up, only to find that I had broken the stem halfway up and wasn't digging deep enough for the actual bulb.

It set in that I was no longer in the sandy loam soils of my East Texas home. I was in the Blackland Prairies, a thick, black, clay soil that cracked during the summer and turned hard as a rock. The soil type ran down from North Texas, through Dallas, and through the central part of the state, ending near Austin. After half an hour of digging, I had about twelve bulbs. This was not going to be an easy crop. I drove back into Dallas that evening and began again the next day. It was the same long story, but it was worth it.

At this point, I still had Ben and Brad with me. We would lay out maps and start visiting each town one by one. Soon we made other discoveries around the state, and all of the towns were located in areas with the same hard, sticky, black soil. My mind wandered back to the first tulips Jimmy Turner from the Dallas Arboretum had given me, and I was very thankful for Jimmy and my two bulbs. I thought about how hard he must have worked to dig them for me.

In March of the next year, I grabbed the whole crew that worked at the Southern Bulb Company at the time and set off with Brad, Ben, and Zac and began the search across Central Texas for the tulip. We were all packed into a little car, with eyes peeled for this botanical treasure. After a long time, we were down in the dumps and ready to give in, when Ben suggested one more turn down a dirt road. A row of old houses appeared beyond thickets of overgrown brush.

Clothes and broken furniture spilled over a screen that must radic dottings of it here and there all over the landscape. We do know though that the bulb propagates quite readily when it is in a spot that it likes.

My travels eventually took me to Georgetown, where the initial patch was sighted in the 1950s. Bill had gone back to the same spot in the meantime as well. The six-foot-diameter patch from the early 1970s had now turned into a twenty-foot patch. The house had changed ownership several times, but the tulip didn't seem to notice. It was happy in the thick, dark soils of the lawn where it did not receive any irrigation.

We still have some *T. praecox* growing on the farm in crates to keep the voles out; it is important for me to see the tulips and continue to work with them. As I see them blooming on my travels, I often think back to how much the red tulip has defined my personal life and the Southern Bulb Company business life and how they have been a small example of some of the wider issues we face with bulbs.

Brad on the farm with the Texas tulips, not long before he proposed to Katie. It still kills me that voles ate most of what you see behind him. Those were some of the hardest digs of our experiences. CLW

Finding a perennial red tulip was a very satisfying moment in my life. The business had almost entirely been started to answer the need in gardening for a warm-climate perennial tulip. With the bulb now found, I am reminded that it is often the anticipation of things that is more exciting than the real deal.

The tulip remains a focus of the business (although we're only building numbers now and not selling it at the moment), but we have also enjoyed some of the other species tulips that have been perennials for us. A large group of species tulips, *T. clusiana*, are perennials for most of the South. This includes 'Lady Jane,' 'Candlestick,' 'Tubergen's Gem,' and 'Tinka.' *Tulipa bakeri* 'Lilac Wonder' also does well, along with some other varieties. Around mid-March, these tulips all begin to open up, and our rows of crates containing the bulbs look like a candy store.

The Southern Bulb Company continued to evolve following the momentous red tulip digs and continues to grow in ways we never imagined. In the end, we found the red tulip, but the red tulip was not the answer. The answer lay in the hunt itself. I am very thankful for the experiences that happened and continue

There is a narcissus immediately in the front, but behind it is the 'Lady Jane' tulip, and in the background is 'Tinka' and more 'Lady Jane' tulips. These have come back and multiplied for five years now on our farm. CLW

At our wedding from left to right: Mickey Wiesinger (my dad), me, John Wiesinger (brother), and Brad Gaultney. CLW

to happen as I live and work as the Bulb Hunter, but I am most thankful for the winding back road that led me straight to my newest partner in the company, Rebecca Joy.

On June 5, 2010, I found myself in a library at the Woodlands Presbyterian Church in a gray suit gifted to me by Bill. A flower pinned to my suit smelled sweet because it was a gardenia picked fresh from the garden of one of Rebecca's coworkers in the flower shop. I stood waiting in the library with my brother and my dad.

We were married indoors with Jacob Shalley, Chad Jones, and Brad Gaultney standing with me. That evening, we danced at a beautiful park pavilion late into the evening, with southern summer sweat running down over our huge smiles. My Corps of Cadets and Texas A&M family circled and sang our school fight song. Rebecca and I danced our final slow song and retired with sounds of laughter and dancing continuing as we left.

What can I say, guys? The cabin now has tile floors, lace curtains, a real hot-water heater, and I'm loving every moment of it with Rebecca. CLW

With the sounds of joy from our wedding just two days old, Rebecca and I left The Woodlands for the cabin. I opened the door to the cabin, and being old-fashioned, lifted her into my arms and carried her across the threshold. Kelly, the sweet potato farmer, had made major improvements in our absence. The cabin was empty with a brand-new tile floor. Against the wall in the main room was a new electrical box, allowing the large toaster oven to be used without tripping a circuit. The bathroom had a full-size water heater in it. The little hallway that ran by the bathroom door and to the back door had been closed off and was now simply a door leading into a slightly larger bathroom. No leaks. To my left was a new door that led to the deck and lake. It was clean, and usable, and comfortable. I looked at Rebecca. We were no longer simply at "the cabin." This was home, and we were together, and happy.

The Bulb Hunter's Bulbs

BY WILLIAM C. WELCH

From the time I met Chris, when he was a student at Texas A&M, it was evident that he had a passion for bulbs, and I delighted in sharing names and places across the South where he could learn about them firsthand as he developed the Southern Bulb Company. Now, we hope you are ready to create your own adventures as you integrate and understand the role bulbs can play in your garden. Following is a concise look at the bulbs as well as some other perennial companion plants to take with you directly to your garden so that you can have distinctive and sustainable color combinations in the years ahead.

A Primer

Heirloom bulbs have become an integral part of our sense of place in the South. As shown by the Bulb Hunter story, Chris and several other pioneers in the heirloom bulb business are now making a wider choice of authentic historical bulbs available for all of us to enjoy. Every season has opportunities for making these easy-to-grow perennials welcome guests for you, your family, and friends. Some will grow almost anywhere, while others respond best to specific sun, shade, and soil types.

The bulbs of oxblood lilies (Rhodophiala bifida) are black with long necks. CLW

Oxblood lilies (Rhodophiala bifida) appear suddenly in late summer/ early fall. They closely resemble amaryllis, to which they are related. CLW

Late Summer and Fall

With our long, warm, and often dry summers, we naturally look forward to cooler weather. Even before this change is obvious, some of the fall-blooming bulbs begin "underground activity" by sending out fresh white roots and taking up moisture from late-summer and early-fall rains. This is triggered by the beginning of shorter days as well as available moisture. The bulbs described here are uniquely adapted to natural growing conditions in the South and require little, if any, artificial irrigation. In fact, they *like* the baking sun and dryness typical of southern summers. (Like Chris, they indulge in long naps during the summer. Chris quickly points out, however, that neither bulbs nor people should be fully exposed to the hot afternoon summer sun.)

This is also a good time to dig and divide, an ages-old form of "cloning" that provides exact copies of the originals. It is an easy way to increase the number of plants while sharing with friends and family. Let's explore the possibilities!

Oxblood Lily, Schoolhouse Lily

Rhodophiala bifida (red)

Natives of Argentina, oxblood lilies have now naturalized over much of Texas and Louisiana and are among the easiest of all bulbs to grow. Their red, trumpet-shaped flowers resemble those of their amaryllis relatives (old-fashioned, hardy red amaryllis bloom mostly in spring) but are about half their size and about 12

Chris in a mass of red spider lilies (Lycoris radiata). *CLW*

inches tall. Chris says that I am too quick to criticize the shortness of the oxblood lily season. He thinks it is not only a great
bulb but also an important reminder that summer is almost over
and welcomes any plant that offers such a cheerful message.

Use oxblood lilies for borders, or mass them in a meadow or
under deciduous trees. Oxbloods thrive in a variety of soils and
are happy in sun or partial shade. They are just about indestructible. According to Chris, about the worst thing that will happen
is you will lose flowers the first year after transplanting. For spectacular displays, plant "drifts" of the 1- to 2-inch-diameter bulbs
4 to 6 inches apart. Ideal planting time is when they are dormant
in late spring and summer, but don't hesitate to plant them whenever you can find them.

![Spider lilies in a garden around the base of a large tree]

Spider lilies (Lycoris radiata) *in the Mangham, Louisiana, garden are combined
with daylilies, 'Country Girl' chrysanthemums, and narcissus for several
seasons of color. At this season, spider lilies are all that are in bloom. WCW*

Lycoris incarnata *are striped pink and white and bloom in early fall in Greg Grant's Arcadia, Texas, garden. Photo Greg Grant*

"Naked ladies" (Lycoris squamigera) *create a traffic-stopping display in a northeastern Louisiana garden. WCW*

Close-up of "naked ladies" (Lycoris squamigera), also known as "surprise lilies" and "pink flamingos." They bloom in late July when little else is in flower. WCW

Spider Lily, Surprise Lily

Lycoris radiata (Chinese coral-red)
L. incarnata (peachy-pink and white)
L. squamigera (bright pink funnels)

Quite a few *Lycoris* thrive for us. One is easily found (*L. radiata*) and is known as "spider lily" (not to be confused with *Hymenocallis*, an early-summer bloomer); another, *L. squamigera*, is called "surprise lily" or "naked lady." There is even a yellow, almost gold-colored specimen found along the Gulf Coast named *L. aurea*.

These all emerge suddenly from late summer to early fall on bare stems that are about 18 inches tall and grow very rapidly once their cycle begins. You'll find *L. radiata* to be the most commonly available and adapted to the widest variety of growing conditions across the South, but as long as you understand the

Bulb mixtures provide several seasons of color under giant pecan trees in the Welchs' Louisiana garden. Early spring features Narcissus tazetta 'Grand Primo,' N. t. italicus, and N. × odorus 'Campernelle,' followed by old-fashioned orange daylilies (Hemerocallis fulva) and surprise lilies (Lycoris squamigera) in July and red spider lilies (L. radiata) in the fall. WCW

Red spider lilies (Lycoris radiata) *adapt and increase in southern gardens from* zone 9 north. CLW

requirements of some of the others, don't overlook them. They are truly spectacular where they perform well.

While touring gardens in Monroe, Louisiana, I noticed a gardening friend out working in her *potager* (vegetable garden) adjacent to the lake. She shared some seeds from a particularly nice purple coneflower (*Echinacea*) and asked if I had seen Chris lately. She said, "Please remind him that I have lots of red spider lilies for him to dig and that the guest suite over the garage will be ready for his visit." I expect Chris will be there shovel in hand. He loves Ann's guest house, a self-contained little bungalow and just the respite for a weary bulb hunter. He said it also helped that the spider lilies she wanted him to dig were down the stairs to the right only twenty steps.

Although the plants are nearly indestructible, the bloom cycle sometimes requires a year or two to resume after transplanting. Sun or partial shade both work well. *Lycoris* will increase fairly rapidly and can be used in large masses or drifts. You can also combine them with daylilies, heirloom daffodils, and other summer perennials and annuals for a continuous display throughout the year. One of Chris's favorite combinations is to interplant them with old-fashioned, single, orange daylilies, 'Grand Primo' narcissus, and snowflakes (*Leucojum aestivum*). We have to give credit for this idea to Jessie Lee Harris in Mangham, Louisiana, who loved to dig up bulbs from the original family farm in the country and spread them about.

Autumn Daffodil

Sternbergia lutea (yellow)

Sometimes called the "fall crocus," these fall bloomers have been a challenge for me to grow. Some historians believe they are the "Lily of the Field" referred to in the Bible. Although I have been given starts of them from the Brenham, Texas, area, they have never prospered in my garden. The bright yellow crocuslike flowers are about 5 to 6 inches tall and appear suddenly in September. They like well-drained soil and are a bright and cheerful addition to the fall landscape. They are thriving for Chris at his farm near Tyler, Texas, and I see nice plantings in the Dallas area. I would like to report some personal success with them someday.

Autumn daffodils (**Sternbergia lutea**) *have naturalized in this Cedar Hill, Texas, lawn just south of Dallas. Their crocuslike blossoms appear in September. CLW*

Close-up of Sternbergia lutea. *CLW*

Perennials to combine with late-summer and fall bulbs

Mexican bush sage (*Salvia leucantha*) makes 3- to 4-foot rounded masses topped with purple or white-and-purple spikes primarily in late summer and fall. It can be used along with autumn aster (*Aster oblongifolius*), with its 3- to 4-foot mounds of lilac-blue flowers with bright yellow centers, in combination with narcissus. About the time the Mexican bush sage and asters are finished, they can be cut back severely to allow light and space for the spring-blooming narcissus or snowflakes.

Lantana comes in trailing or mounding types that can grow 3 to 5 feet tall and wide and come in pink, white, red, orange, or multicolored rounded flower clusters. They prosper in the heat and perform well in sun or partial shade. Little or no irrigation is needed once plants are established.

Garden mums can be a nice addition to the landscape. Recycling florist mums into the garden can be done with some success, but garden mums are developed especially for garden display. 'Country Girl' is a good example with its masses of bright pink, daisylike displays that start in mid-October. It is also known as 'Ryan Gainey,' named for the well-known Atlanta garden designer. To keep them compact, prune them back a couple of times in summer and just above the ground after the blooms fade and frost has occurred. I enjoy some of the old, small-flowering, yellow and rusty-red pom pom–type mums grown by my grandmother and aunts in Yoakum, Texas. I also like masses of 'Country Girl' mums interplanted with 'Marie Daly' roses and bordered with my "recovering" white rain lilies in College Station. Mums should be divided early in spring or rooted from cuttings in early summer. Chris does acknowledge that I have a nice display of mums but thinks they would look better if I would contain them more and allow his beautiful (re)planting of white rain lilies to grow up among them.

Rubeckia × 'Herbstonne' blooms for about two months in midsummer. The pink flowers climbing in it are coral vine (Antigonon leptopus), *which have pink or white heart-shaped flowers in late summer and fall. WCW*

Autumn aster (Aster oblongifolius) *is a low-maintenance, mounding, long-lived perennial.* WCW

Winter and Early Spring

Our relatively short cold period in the South is the major reason so many of the "mainstream" daffodils, hyacinths, and tulips are annuals for us. They require more hours of chilling temperature (total number of hours below 45 degrees F) to satisfy their rest or dormancy needs. For this reason it is so important to notice, identify, and verify the bulbs that keep coming back and plant them today in our gardens. They have proven to be drought tolerant, insect and disease resistant, as well as beautiful and often fragrant.

Bulbs that flower in late winter and early spring begin to emerge as early as October or November. Fresh green foliage that is rarely damaged by freezing weather is followed soon by the first buds and flowers of early-blooming narcissus and snowflakes. It is a bit confusing, but daffodils and narcissus are really the same thing. Smaller-flowering forms are referred to as narcissus or paperwhites, and the larger trumpet types as daffodils. Some paperwhites can flower as early as Thanksgiving, but trying to sort these very early bloomers out at the commercial growers is a bit of a challenge.

This garden hybrid chrysanthemum has masses of old-fashioned daisylike blooms in the fall. WCW

The lake at the cabin near Mineola, Texas, has provided Chris and his guests food and pleasure as they share bulb-hunting experiences. WCW

Flowering quince (Chaenomeles lagenaria) *is a tough and useful shrub that blooms in January or February when little else is blooming.* CLW

Narcissus

Narcissus orientalis (white with gold cups)
N. tazetta italicus (white with small yellow cups)
N. t. 'Grand Primo' (white petals, pale yellow centers)
N. jonquilla (solid gold)
N. × odorus (yellow)
N. pseudonarcissus (bright yellow)
N. 'Golden Dawn' (yellow and orange)

Chris and I agree about the value of January-blooming narcissus, like Chinese sacred lily (*N. t. orientalis*), in its double- or single-flowering types, and *N. t. italicus*, which performs really well as far south as the Gulf Coast. They not only look good in the garden but are great as cut flowers in the home.

Island beds of bulbs and other perennials begin the color season with narcissus and snowflakes (Leucojum aestivum) in late February and early March. WCW

Chinese sacred lilies (Narcissus tazetta orientalis) *start in zone 8b and thrive through zone 10 and are available in single- or double-flowering form. They are among the earliest of the narcissus to bloom.WCW*

Narcissus tazetta *'Grand Primo' may be the best-adapted narcissus in the South. They bloom prolifically in February and March with nicely scented clusters of white flowers and creamy-yellow centers shown here in a Wedgwood "posyholder." WCW*

Narcissus tazetta italicus *bloom on a historic site in Natchitoches, Louisiana.*
CLW

Narcissus tazetta italicus *bloom early and naturalize well, even in coastal areas of the Deep South. CLW*

'Erlicheer' narcissus are thought to be a double-flowering form of the revered 'Grand Primo.' WCW

Jonquils (Narcissus jonquilla) *are a favorite for easy-to-grow fragrant flowers. WCW*

A white Lady Banks rose (Rosa banksiae 'Alba Plena') covers a gazebo at Fragilee, creating color and delightful fragrance as the heirloom bulbs, other antique roses, and perennials begin their spring show. WCW

A Word about Landscape Design

Some gardens are designed professionally from the beginning, while others evolve over time with inputs from the family, gardening friends, media sources, and visits to other gardens. All of these methods can result in successful outcomes. The "hardscape" consists of walks, drives, constructed edges, arbors, fences, and paths. Landscape architects are the professionals trained to deal best with these issues, although there are many landscape contractors and consultants who are worth your consideration as well. Think about landscape design as a process to bring *order* into your property.

Of equal importance is maintenance. Think of maintenance as bringing *discipline* into your outdoor spaces. Whether you hire a professional to do this or it becomes a family project, it should reflect your lifestyle and tastes.

Chris suggests that "before planning your landscape, consider moving to a town with lots of college students you can hire to prune your prickly roses, lift heavy objects, spread smelly items around plants, and so on. It has done wonders for Bill's garden."

Bulb Planting 101

The ideal time to plant for the winter and early-spring season is October through December. Thanksgiving is an ideal time to plant narcissus—the turkey goes into the oven and the bulbs go into the garden! In reality, bulbs can be planted most any time, but root growth begins naturally as summer ends and fall rains cool and soften the summer-baked soils.

Although it may seem elementary, first-time gardeners need facts, like which end is up when planting a bulb? The answer is pretty simple. The pointed end goes up, and the rounded one goes down. Roots will emerge from the rounded end while leaves and flowers come from the top. The rounded and larger bottom is known as the bulb plate. If this bulb plate is cut off, the bulb will die, but if even a part of the plate remains, it will usually heal itself and prosper. If ever unsure which end is up, plant a bulb sideways!

Depth of planting is another easily solved issue. Generally, let the pointed top of the bulb be buried about twice its height. In heavy soils plant a little shallower than in sandy ones. Spacing may be a compromise between your budget and your desire for a quick effect. Smaller bulbs may be set as close as 2 to 3 inches apart, while larger ones can be planted as much as a foot or so apart. Large crinums and *Hymenocallis* may be set as much as 2 feet or more apart. Fertilizer is best used sparingly, but organic material such as composted pine bark or compost from your own garden when worked into the top 8 to 10 inches of the soil is usually helpful. Organic fertilizer sources such as cottonseed or alfalfa meal applied at 5 pounds per 100 square feet of bed area are helpful, but high-nitrogen chemical fertilizers are not recommended. It is a good idea to water the bulbs as soon as they are planted and occasionally during dry spells the first season after planting.

Jonquils (Narcissus jonquilla) *could tell stories from the past of this old homesite near Crockett, Texas. In addition to multiplying freely from offsets, this jewel of the garden reseeds as well. WCW*

By mid-February what I consider the best of all narcissus for the South begin their show. The vigor of *N. t.* 'Grand Primo' is unrivaled across the South. Three or four stems per bulb emerge over a one-month period. Pleasantly fragrant clusters of ivory-white blossoms having slightly darker cups grow like they are on steroids! These are closely followed by another southern treasure in the form of jonquils, which are sometimes referred to as "sweeties" because of their delightful fragrance. At least two forms, both bright yellow, are abundant in gardens of the old South. *Narcissus jonquilla* blooms first and multiplies by offsets as well as from seed. *Narcissus intermedius* (Texas stars) have

Hoop petticoat narcissus (Narcissus bulbocodium) have unusually shaped flowers and are old southern favorites. CLW

A drawing by artist Ann Swan reproduces the subtle beauty of the dormant bulb. CLW

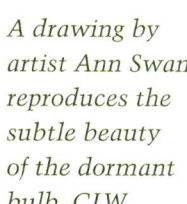

slightly taller foliage with fragrant blooms nestled in the foliage.

Yet another star performer is *N. × odorus*, the 'Campernelle,' which along with 'Grand Primo' and the jonquils, is among the most elegant and useful bulbs ever introduced to the South. Its bright yellow trumpets are intermediate in size between those of jonquils and daffodils, and the reedlike leaves are attractive as well. Midspring ushers in yet another jewel for our gardens *N. pseudonarcissus*, which looks very much like a typical, medium-size daffodil. Perhaps not quite as vigorous as the others mentioned, it has an elegance all its own.

Although of more recent origins, *N.* 'Golden Dawn' has been shown to prosper and increase over most of the South. Its vigor is close to that of 'Grand Primo,' and the clusters of yellow-and-orange flowers make a real show in the garden. Narcissus, along with *Leucojum*, certain species tulips, Roman and grape hyacinths,

Narcissus × odorus 'Campernelle' is one of the very best heirloom narcissus for most of the South. *Attractive, reedlike foliage appears in late fall with the bright yellow, elegantly shaped flowers occurring over several weeks in mid- to late winter. A delightful fragrance is another virtue. WCW*

Lent lily (Narcissus pseudonarcissus) has been popular in the South for many generations. WCW

Narcissus *'Sir Winston Churchill' is one of the latest-blooming narcissus for the South. Although they thrive in hardiness zone 8, there are better choices for areas closer to the Gulf Coast. WCW*

Narcissus 'Golden Dawn' in healthy clumps against a white picket fence in northeastern Louisiana. Blooms come late in the season, usually in late March and April. It is always in my top-five list of narcissus for the South. Clusters of bright yellow flowers with orange centers appear in great abundance. It is best adapted to zone 9 and north. WCW

Close-up of Narcissus 'Golden Dawn.' WCW

and *Ipheion*, can create at least three months of color. 'Golden Dawn' is followed by *N.* 'Sir Winston Churchill' (fairly large flowering clusters of white with yellow centers) extending into April.

Chris enjoys seeing 'Golden Dawn' in our Mangham, Louisiana, garden: "It seems that every March I manage to find myself in a situation where I need a home near Monroe. I like the way the 'Golden Dawn' have been scattered around borders and the way they make a nice splash of color as the narcissus finish their blooms. I always manage to raid the fridge, garden for bulbs, or collect pies from the Welch's old family friends nearby."

Snowflake

Leucojum aestivum (white with green dots)

Just behind the early white and yellow narcissus are snowflakes (*L. aestivum*), one of the best bulbs for the entire South. They usually start blooming in February and continue for at least six

Leucojum 'Gravetye Giant' (snowflakes) have larger flowers than the species type and are just as hardy and prolific. WCW

Snowflakes (Leucojum aestivum) are interplanted among azaleas in a Houston garden. WCW

weeks. The species form has been around for at least two hundred years but has been joined by the cultivated 'Gravetye Giant,' named in honor of the home of famous English horticulturist William Robinson (1838–1935). The lily-of-the-valley-like flowers have a dot of green on each petal and are almost twice the flower size of the species form. Both are great for our southern gardens. All of these bulbs prefer sunny locations but will also naturalize and increase under deciduous trees like pecans and oaks. Since they love being dry in summer, they are often found growing in the root flares at the base of tree trunks. *Leucojum* grow in almost any soil and have attractive foliage and flowers that range from 12 to 18 inches tall.

"Minor Bulbs": Roman Hyacinth, Grape Hyacinth, and Starflower

Hyacinthus orientalis (blue, white, and rarely pink)
Muscari neglectum (purple)
Ipheion uniflorum (white to dark blue)
Narcissus are the star performers for most winter and early-spring gardens, but some of the "minor" bulbs like Roman and grape hyacinths, species tulips, and spring starflower (*Ipheion*) can add

Blue Roman hyacinths (Hyacinthus orientalis) *are among the most fragrant and charming flower bulbs grown in the South. CLW*

Pink Roman hyacinths (Hyacinthus orientalis) *in an antique glass vase. My original bulbs were from Elizabeth Lawrence's garden in Charlotte, North Carolina. Photo Cynthia Mueller*

interest and color. The fragrance and simple beauty of Roman hyacinths (*H. orientalis*) in blue, white, and rarely pink are not excelled. Once you experience Roman hyacinth fragrance, you will always remember it. The white form usually blooms early in January, followed by the blues and pinks. Dear friend and gardener extraordinaire, the late Flora Ann Bynum from Old Salem, North Carolina, along with author and friend Greg Grant from Nacogdoches, Texas, were once referred to as the "Roman hyacinth police" and often asked to judge the authenticity of various hyacinths found throughout the South. Flora Ann recalled that Roman hyacinth bulbs and cut flowers were often sold in flower and farmers' markets in North Carolina. Producing them commercially has been frustrating for Chris and others because they

Pink Roman hyacinths (Hyacinthus orientalis) *are combined here with the succulent* Sedum potosinum. *Both like relatively dry conditions in the summer and seem to be compatible. WCW*

Divisions of white Roman hyacinths (Hyacinthus orientalis) *have just been dug and separated. This can be done at any time, but when dug while actively growing, the plants should be quickly reset and watered.*

Grape hyacinths (Muscari neglectum) *will naturalize in lawns or other areas with good drainage. CLW*

Spring star-flowers (Ipheion uniflorum) *rebloom during February and March and thrive on well-drained soils. WCW*

increase slowly and are a favorite of moles, which devour them as quickly as they are planted. Roman hyacinths are rarely more than 8 to 10 inches tall and prefer well-drained soils and at least a half day of direct sunlight. Irrigation is not required.

Grape hyacinths (*M. neglectum*) can sometimes be found in old cemeteries and homesites. When profuse, they almost challenge Texas bluebonnets for color and show. Their 4- to 6-inch dark blue spikes like sunny locations and well-drained soils. Sometimes known as "starflower," *I. uniflorum* ranges in color from nearly white to dark blue. They are another small plant and flower, usually less than 6 inches tall, with flat, grayish-blue leaves that reveal their kinship to garlic when bruised or crushed. They, too, like sunny, well-drained sites and love "living on the edge" of my gravel driveway, making a pretty picture each spring at the foot of a large 'Cl. Red China' rose.

Iris

Iris albicans (white)

A striking iris distinguishes itself in southern gardens in early spring. Many areas of the South can grow tall bearded irises as well as Louisiana irises, but *I. albicans* is the toughest of them all. Often referred to as "cemetery whites," their grayish-green fans of foliage are evergreen and thrive on neglect. They prefer a sunny, well-drained location and little or no supplemental irrigation.

I was invited to collect a few *I. albicans* from a very old site in Camden, South Carolina, while attending the Southern Garden History Society meeting. The site dated back to George Washington's time. A native of Yemen, *I. albicans* thrives from the northern parts of the southern states to the Gulf Coast. Yet commercial sources are almost nonexistent. But I think I may have made an impression on Chris. Last spring, he noticed a patch of cemetery white iris blooming in a ground cover of 'Texas Appleblossom' and 'Homestead Purple' verbena in our Mangham garden, and he was finally hooked. Chris continues to work with a propagator on these and is having much success bringing them to market.

Cemetery white irises (Iris albicans) *thrive on the Gulf Coast to the most northern parts of the South. WCW*

Tulip

***Tulipa praecox* (red)**
***T. clusiana* (red and white)**
***T. chrysantha* (yellow and red)**

Naturalizing tulips in the South continues to be a challenge. By "naturalizing" I mean adapting tulips to the relatively short winters and inconsistent moisture levels across the South. The most provocative glimpse of hope seems to be from *T. praecox*, which Chris has dubbed "Texas tulips." Chris and I call them by different names, but I found them first! This is the tulip I first encountered at Southwestern University in Georgetown, Texas, where I observed a small patch of large-flowering, bright red tulips with bluish-green leaves blooming in the St. Augustinegrass lawn of a prairie cottage across the street from the campus.

"Georgetown tulips" (Tulipa praecox) *are a promising possibility for hardiness zones 7 and 8. WCW*

Chris collected some from this site and others he found and planted them at the Southern Bulb Company farm, where he has generally been pleased with their performance, although some of his customers have had better success than others in growing them. "Customers have definitely let me know about this. It has been a challenge to offer the right bulbs for the right consumers. At the moment, we are not selling the Texas tulip, but we hope to have them in the future." Texas tulips prefer sunny, well-drained sites and don't mind heavy clay. They have not done well as far south as Houston but usually prosper in the Waco and Austin areas and north and west of there. They like to dry out in summer and prefer non-irrigated areas.

'Lady Jane' tulip (*T. clusiana*) has also enjoyed some success in the South. These are small-flowering, graceful plants and about half the size of a typical garden tulip, with red exterior petals and white inner ones, resulting in a gay, candy-striped effect. They are attractive in masses and demonstrate "sleep movement" by expanding to flat white stars at noon but closing to tight red buds each evening. The species *T. chrysantha* is similar but blooms a couple of weeks later and has yellow-and-red coloration. 'Cynthia' is a pale yellow hybrid of the two, and others include 'Tinka' and 'Tubergen's Gem.'

Tulips may be dug and divided as their leaves yellow and brown in spring. The ones mentioned here do not require refrigeration and do best when left in the ground. If they are thriving for you, dividing may be done every three or four years.

The owner of the original patch in Georgetown graciously shared some with me, and I am eager to see how they are going to perform. They continue to slowly increase and return each year in my garden.

Perennials to Combine with Late-Winter and Early-Spring Bulbs

Phlox have always been a personal favorite, and when my friend Ruth Knopf from Sullivan's Island, South Carolina, gave me a start of *Phlox pilosa* she had received from Elizabeth Lawrence, I could not have been more pleased. They thrived at Fragilee, my home in Washington County, Texas, and when Jesse Lee Harris planted them at our Mangham, Louisiana, place, they created a lot of attention. Within two years of first setting them out, he received many requests to share them with gardening neighbors.

Commonly known as "prairie phlox," *P. pilosa* is native to the South. Elizabeth Lawrence exchanged plants with gardening friends all over the South, and we don't know where she got her start. We do know that she collected plants and wrote about them in her column in the *Charlotte Observer* for many years.

Other spring-blooming phlox are also great companions for spring bulbs. Moss pink or thrift (*P. subulata*) is a neat, mosslike ground cover that starts its display in early spring with sheets of bright pink, white, or purple. It's a bit fussy about drainage, doing best in sandy soils. Louisiana phlox (*P. divaricata*) is another favorite. Masses and borders of these 1-foot-tall, beautiful violet-purple flowers prefer partial shade and slightly acidic soils. Spring-blooming phlox are best planted or divided in the fall. It is amazing to watch them develop over winter and at times dominate the garden by April. It is not unusual for all three of these phlox to bloom for three months in spring.

Although I have always been fascinated with columbines, my interest increased when I learned about our native Texas species. I first saw *Aquilegia canadensis* in about 1965 in Lynn Lowrey's Houston nursery. It has reddish and yellow flowers atop attractive blue-gray foliage, which seemed exotic and unusual. Native to streambeds in Central and West Texas, they have another asset in their willingness to bloom in partially shaded areas. A few years later I saw the native Texas yellow form (*A. chrysantha hinckleyana* 'Texas Gold') and immediately felt that it was a

Prairie phlox (Phlox pilosa) were started from a few divisions and have spread to make a major display in the Welch Louisiana garden. WCW

winner. Native to the banks of tiny streams in far West Texas, it is sometimes sold as 'Texas Gold.' Healthy, 18- to 24-inch mounds of blue-gray foliage are topped by many stems of bright golden flowers for a month or more each spring. A dilemma occurs when both these columbines are planted nearby. Promiscuous behavior between them results in seedlings coming up that have characteristics of both plants. Chris doesn't allow promiscuous behavior in his garden and has decided to plant only the more showy 'Texas Gold' form. We'll hope that nearby neighbors don't plant *A. canadensis* close enough for an accident to occur. More easily available cultivated types of columbines are popular in the South, but these two native Texas species are more heat resistant and likely to be perennials.

Perennial forms of verbenas can add lots of color in late winter and spring. The low-growing viny character can provide sheets of flowers that combine well with bulbs. Although usually considered perennials, verbenas tend to die out in their centers over time. I particularly like 'Texas Appleblossom' and 'Homestead Purple' used together with 'Grand Primo' or other good narcissus.

Chinese ground orchid (*Bletilla striata*) is one of the few terrestrial orchids suitable for general cultivation in Texas and the Gulf South. It thrives in the shaded or partially sunny garden with a moist, humus-rich soil. Spikes of purplish-pink flowers emerge in early spring and may be damaged by late-spring freezes unless mulched annually. Originally from China and Japan, Chinese ground orchids were first cultivated in England around 1994. When established, almost every shoot will contain up to fourteen individual, orchidlike flowers. Foliage is 12 to 18 inches and pleated. Established clumps can have dozens of flower spikes, with individual flowers resembling a miniature Cattleya orchid. Propagation is by dividing large clumps in late summer to early fall.

Late Spring and Summer

Our gardens in the South are at their most colorful in late spring, and we always wish that we had planted more! The "extra" space we noticed earlier has all been filled. Cool-season annuals like

The front gate planting at Fragilee in Washington County, Texas, includes the old Johnson's amaryllis (Hippeastrum × johnsonii) *rescued from the original homesite of the house. 'Cl. Red China' rose, Chinese sacred lily* (Narcissus tazetta orientalis), *and 'Grand Primo' narcissus along with 'Henry Duelberg' salvia provide many months of low-maintenance color. WCW*

The Chinese ground orchid (Bletilla striata) is a magenta-colored hardy orchid that is well suited as a perennial to be planted and left in outside gardens across the South. WCW

old-fashioned petunias, violas, poppies, sweet alyssum, sweet peas, larkspur, nicotiana, and poppies not only create garden pictures but scent the garden with a blend of wonderful fragrances. Southern favorite shrubs like sweet olive (*Osmanthus fragrans*), banana shrub (*Michelia figo*), and the last blossoms of winter honeysuckle (*Lonicera fragrantissima*) blend with sweet shrub (*Calycanthus floridus*) and citrus blossoms to add further complexity to the air.

As we move into the hotter part of summer, treasures unique to warm climates come into bloom. Flowers both large and small surprise gardeners with glorious colors. Late spring is the time for some of our favorite bulbs and one our favorite drinks to help keep us cool!

Hardy Amaryllis

Hippeastrum × johnsonii (red and white)

The all-time favorite amaryllis for the South, *H. × johnsonii*, is sometimes known as "St. Joseph's lily" or "Johnson's amaryllis," after its originator, who crossed two species of amaryllis around 1790 in England. No other amaryllis is as vigorous, and its combination of red flowers, white stripes, and delightful fragrance make it a southern favorite. Dividing the clumps will encourage more plants, but they don't seem to mind being crowded. Commercial availability is a challenge because they are smaller bulbs and sometimes unknowingly perceived as inferior bulbs, but generous gardeners in the know have shared them with one another for more than two centuries.

When we moved the old house we call Fragilee a few hundred yards to its new location, I examined the old site looking for

Johnson's amaryllis (Hippeastrum × johnsonii) are surviving and flourishing in an abandoned Independence, Texas, garden. WCW

Back gate at Fragilee, a Washington County weekend home, with 'Cl. Red China' rose, 'Cl. Cramoisi Superieur' and hardy red amaryllis salvaged from the original site. WCW

evidence of former plantings. The house had been abandoned for about twenty-five years, and cattle had been browsing the area. I did find a few leaves of what looked like amaryllis, dug out eight or ten bulbs, and moved them to a raised planter at the left side of the gate. There were also a few narcissus that I later identified as *N. orientalis*. The amaryllis multiplied quickly and bloomed the next year after planting.

The plantings on both sides of the gate have evolved over ten years to survive dry conditions and alkaline soils. Soon after starting his bulb farm, Chris was overnighting at Fragilee and offered to divide and reset my amaryllis. He also asked if, "while he was there," I would share my Chinese sacred lilies. He dug them late one evening, using his headlights for light.

I have been pleased with the progress of this combination, even with the "temporary setbacks" from the Bulb Hunter. The dryness of the area is alleviated a bit by several sprinkler heads, but I believe the groupings would survive now with no outside irrigation. An old red China rose is nestled into the Johnson's amaryllis.

Long favorites as holiday season gifts, modern amaryllis have larger, flatter, trumpet-shaped blossoms as well several scapes (stems) per bulb. Newer cultivars like 'Apple Blossom' (pink and white), as well as pure white, striped, and solid reds and oranges, continue to increase in popularity. Cold protection is recommended north of zone 8. Amaryllis are good in pots and don't mind being crowded. Use them in masses in full sun or partial shade.

Gladiolus

Gladiolus byzantinus (magenta)
G. dalenii (orange)
Byzantine gladiolus are enjoying renewed popularity throughout the South. About half the size of conventional gladiolus, this hardy form is a welcome addition to both old and new gardens. Unlike those of modern hybrids, Byzantine corms (they are not in a strict sense bulbs) do not require staking or annual digging and storing. Originally from Turkey, they prosper in our heat and dryness. It is best to purchase them from a reliable source since some of the ones in the market don't have the vigor of our time-tested southern heirlooms. Chris is quick to point out this difference as he explains why his Byzantines cost dollars per corm and the others cost only pennies per corm. Masses of their spiky flowers used in perennial borders are really beautiful for several weeks in late spring. Divide after they die back in early summer. Gophers seem to love all gladiolus (even those Chris planted in plastic crates buried in the field).

Another interesting old gladiolus is the parrot glad (*G. dalenii*) with striking hooded flowers on tall spikes of orange and greenish-yellow. Parrot glads are 3 to 4 feet tall and prefer well-drained soils and full sun or partial shade. Gladiolus are especially effective behind masses of lower-growing Louisiana phlox (*P. divaricata*) or prairie phlox (*P. pilosa*).

Amaryllis (Hippeastrum), *Byzantine gladiolus* (Gladiolus byzantinus), *and old-fashioned petunias provide nostalgic glimpses into gardens of the past. WCW*

Byzantine gladiolus (Gladiolus byzantinus) *bloom along with early bursts of roses and prairie phlox* (Phlox pilosa). *WCW*

Parrot gladiolus (Gladiolus dalenii) *has large, striking coral- and chartreuse-striped flowers. WCW*

Easter Lily

Lilium longiflorum (white)

Cold hardy in most of zone 8, *L. longiflorum* provides elegant, white, funnel-shaped flowers each year in May. If you have access to heirloom Easter lilies, you will probably find them more vigorous than recycled ones from the florist. The rosettes of pointy, green foliage are an attractive background for the beautiful flowers. Easter lilies prefer a sunny, well-drained site and reach 2 to 3 feet tall, requiring little, if any supplemental irrigation.

Tiger Lily

Lilium lancifolium (orange)

Tiger lilies (*L. lancifolium*), with their orange trumpets, are among the most reliable lilies for southern gardens. They can be divided every three or four years, or you can plant the bulbils (miniature bulbs) that appear in the axils of the leaves next to the stems. They will usually bloom the second year after plant-

Easter lilies (Lilium longiflorum) continue to be a good perennial choice in hardiness zones 8–10. The ones shown here are from Pam Puryear's grandmother's garden in Navasota, Texas. WCW

ing. Tiger lilies prosper in our Mangham garden where they are in filtered shade. I see them doing well in sunnier locations as well, and they usually reach 2 to 3 feet tall. Little or no irrigation is required.

Tiger lily (Lilium lancifolium) *is probably the best-adapted garden lily for the South. Its bright orange, funnel-shaped flowers are dotted with black. Bulbils are tiny bulblike structures formed where the leaves meet the stems. When planted, these can produce blooming-size plants in two years. CLW*

Crinum

Crinum jagus (white)
C. powellii 'Album' (white)
C. × **herbertii** (milk and wine)
C. 'Mrs. James Hendry' (blush)
C. 'Carroll Abbott' (white and burgundy stripes)
C. 'Ellen Bosanquet' (dark pink)

Sometimes referred to as "milk and wine lilies" or "ditch lilies," these venerable bulbs range in color from pure white to dark pink and also have many striped forms. Some crinums have straggly foliage, while others are neater in appearance. They are among the easiest garden plants to grow, and some rebloom numerous times through our long, hot growing season beginning in midspring and ending at first frost.

Among the most beautiful is a pure white species form (*C. jagus*) that has beautiful dark green, strap-shaped foliage. These are only

'Bradley' is a particularly nice crinum. It reblooms throughout the growing season and has more attractive foliage than most crinums. WCW

cold hardy from zone 8 south and usually bloom for several weeks in late May and June. Flowers are tulip shaped and smell like vanilla.

I introduced Chris to this plant in its narrow-leaved form: "Bill told me the story of the time he had first seen it in Wayne Womack's Baton Rouge garden. He described it as a mounded hill in the back part of the landscape filled with these blooming bulbs and their vanilla scent creating a spectacular experience. A few years ago the site had been abandoned by the previous owners and was covered in overgrown brush. The bulbs were surviving, but in an area where plants grow so quickly and vivaciously, I imagine it will not be long before they are choked out."

Many outstanding crinums grow in a wide variety of soils, including heavy clay. I was given a large collection by noted horticulturist Scott Ogden a number of years ago and have become attached to some. My favorite is the pale pink 'Mrs. James Hendry,' which blooms from early spring until frost with wonderfully scented flowers that are outstanding both in the garden and as cut

*Close-up of
Crinum 'Mrs.
James Hendry.'
WCW*

flowers. Just one stem can scent a room and provide continuing pleasure as each bud opens.

Another choice is *C. powellii* 'Album,' commonly found in old gardens across the South. Another well-known white crinum is Burbank's 'White Queen,' named for the famed California horticulturist Luther Burbank. 'Bradley' is a lovely rose-pink form with good foliage, and 'Ellen Bosanquet' is among the most common and revered. 'Stars and Stripes' is one of Chris's favorites and produces big bouquets of white- and pink-striped flowers. 'Carroll Abbott' is another striped favorite. Crinums are really hard to kill! It can be done if you dig them and cut off the bulb plate, which I have done when digging crinums in very hard clay in summer.

Crinum bulbs can get huge! *Crinum asiaticum* is well adapted from zone 8 south and can be 6 to 7 feet tall with bulbs 8 to 12 inches in diameter. Dividing old clumps can be seriously hard work, especially in the heat of summer.

Rain Lily

Cooperia (yellow, orange, white)
Zephyranthes candida (white)
Z. grandiflora (rose-pink)
Habranthus robustus (pinky-white)

Don't let the scientific names of these tough perennials intimidate you! Although long rooted in the past, these beautiful plants are a great fit for our gardening future. They are part of the group of bulbs known as amaryllids (such as amaryllis and crinums), and rain lilies are well suited to live on natural rainfall even in the semi-arid southwest.

Some rain lilies are native to the South and West where the yellow, orange, and white forms (*Cooperia*) can be quite impressive following summer rain showers. My favorites, however, are the hybrids and species from Argentina, which can be really impressive as soon as three or four days following a summer rain. That they thrive with no irrigation, have no serious insect or disease problems, and provide abundant color in late summer and fall make them ideal.

White rain lilies (Zephyranthes candida) are useful as borders or in clumps. CLW

White rain lilies (Zephyranthes candida) thrive under relatively dry or wet conditions. Their foliage looks good for most of the year, and flowers occur repeatedly in late summer and early fall. WCW

The rain lily *Z. candida* is native to the banks of the Rio Plata in Argentina, where it is known for its spectacular blanket of bloom. The star-shaped, pointed buds of these white, lilylike flowers remind me of crocus. They begin to appear in August and continue into October. Handsome evergreen foliage makes them useful even when not in flower. Blooms are triggered by rain showers in late summer and fall. Unlike most other rain lilies, *Z. candida* will actually thrive partially submerged in water. Chris is growing some commercially this way at the edge of a catch pond on his farm.

I noticed the large, beautiful, dark pink flowers of *Z. grandiflora* when I first began admiring bulbs as a preteen in Houston. They were in the garden of Maxine Wendler, a family friend who lived down the street in Garden Oaks. Ima Hogg also grew them in the Diana Garden at Bayou Bend. She was the daughter of Governor James Hogg and a noted art and furniture collector and Houston philanthropist.

Pink rain lilies (Zephyranthes grandiflora) used as a landscape plant where they usually rebloom numerous times during the spring, summer, and fall. WCW

Another favorite is *Z.* × 'Grandjax,' with its pink and apricot flowers. I got my start from Scott Ogden, who was advocating rain lilies before anyone else I know. Chris spotted some I planted in the parterre garden at Fragilee, which I shared with him and others. After one of the hottest and driest summers on record, I was thrilled to see masses of color appear three times following the first rains of September.

Habranthus robustus may be the best of all the rain lilies and is now commercially available on a limited scale. Chris describes the color as blush with an almost white throat. He has had inconsistent results from seed, but a batch from Cynthia Mueller germinated more than five thousand plants that produced blooming-size bulbs within a year in a greenhouse. Chris speculates that this is great news for the commercial grower but also points out that once customers obtain some, they will never need to buy them again. I have been amazed at how quickly *H. robustus* have multiplied and even reseeded in my garden. They surprised me by doing well in the partial shade of a gazebo covered with yellow Lady Banks roses, blooming three or four times each summer.

Pink rain lilies (Zephyranthes grandiflora) are among the showiest summer perennials. They are native to Argentina but thrive all across the South, where they have been grown for many generations. WCW

Spider Lily

Hymenocallis 'Tropical Giant' (white)
H. liriosme (white)

The spidery white flowers of *H.* 'Tropical Giant' top foliage that can easily reach 3 to 4 feet. *Hymenocallis* will grow in poorly drained soils and make handsome masses of bold foliage. They bloom in June and July and are usually cold hardy in zone 8 and south. Masses of them are found in old gardens of San Antonio and Austin. Much like their crinum cousins, spider lilies are nearly indestructible.

Chris says, "I have a picture of a nice clump of these blooming in Bill's garden right before I left for a Fourth of July celebration. The more I pondered the best way to describe these white

Spider lilies (Hymenocallis 'Tropical Giant') have bold tropical foliage and showy white flowers in early summer. These plants tolerate wet or dry conditions. WCW

A row of Hymenocallis 'Tropical Giant' emerges from 4- to 6-inch-diameter bulbs and tolerate wet or dry conditions, as well as heavy soils. WCW

blooms, the more they reminded me of white fireworks that explode into stars, with their points trailing off, giving them a spectacular, yet droopy appearance."

Spider lily foliage makes a really bold statement in the garden, which Chris and I noted in a public park in Fredericksburg, Texas. Early German settlers in the area quickly realized the convenience of the native stone and wood from native, rot-resistant cedars (*Juniperus ashei*). The park's designer had featured native limestone for paving and walls. The cedar had been used to make a grape arbor, an enticing summer retreat for area visitors coupled with the promise of tasty fruit for wines and jellies.

Large masses of *Hymenocallis* were mirrored by similar ones of *Crinum* 'Ellen Bosanquet.' Island plantings containing several plants of 'Knockout' and 'Belinda's Dream' roses seemed to add just enough color, while the boldness of the crinum and *Hymeno-callis* foliage completed the visual picture.

Perennials to Combine with Late-Spring and Early-Summer Bulbs

The heirloom daylilies (*Hemerocallis fulva*)—single orange, double, and yellow forms—make a big display in early summer. I was first fascinated with the double orange daylily ('Kwanso') as an early teen. The driver of the school bus knew I was into gardening and shared some of her daylilies with me. They did very well in our Houston garden, and so did the single orange one that is the most common daylily grown in America. Chris is fond of the plantings at the trunks of pecan trees in Mangham where they are combined with spider lilies (*Lycoris radiata*). These are enhanced with summer annuals like torenias and coleus, and the bulb *N. t.* 'Grand Primo' in early spring.

Louisiana irises provide striking foliage and beautiful flowers from mid- to late spring. They range in color from white to yellows, purples, blues, bicolors, and rusty reds. Wet or poorly drained soils are not a problem for these plants, but Louisiana irises also thrive in drier locations where organic material has

The old double form of the common orange daylily (Hemerocallis fulva 'Kwanso') has been popular since Victorian times. WCW

been added to the mix. Dividing every two or three years keeps them vigorous, and blooms are more prolific if they get at least a half day of direct sun.

Elephant ears (*Alocasia* and *Colocasia*) have been popular in southern gardens since Victorian times. Members of the genus *Colocasia* prefer wet or boggy soils and can become invasive along water edges. Those in the genus *Alocasia* have more pointed foliage and prefer better-drained soils. Recent new ones like 'Lime Zinger' (chartreuse) and 'Blackie' (dark, almost black) have become very popular in perennial borders. Their giant leaves provide bold textural contrast.

Another more recent addition to the southern plant palette is *Farfugium japonicum*, sometimes called "leopard plant." I tried the conventional form of this plant years ago and found it lacked vigor in our hot climate and barely survived. But some of the newer cultivars make a really bold foliage statement in shaded or partially shaded gardens with their 6-inch diameter and 3- to

Two clumps of Narcissus *'Grand Primo' are thriving under the partial shade of deciduous trees. CLW*

Hibiscus moscheutos × 'Lady Baltimore.' The owner counted thirty-five open blossoms on a day in late July. WCW

Yellow bells or esperanza (*Tecoma stans*) is only limited by its inability to withstand cold. 'Gold Star' is a short, early-blooming selection by Greg Grant and is widely used in the horticulture trade. It is normally root hardy through zone 8 and grown in containers farther north. It is amazingly heat and drought hardy and thrives all the way to the Gulf Coast. Flowers are most prolific during July and August when temperatures are at their peaks. The 12- to 15-inch clusters of 3- to 4-inch bell-shaped flowers are a really bright yellow. In areas where frost occurs, cut them back during late winter near ground level. If they don't freeze in your area, prune and shape them in late winter. Esperanza does not perform well in the shade. It appears that more sun and heat produce more flowers. Esperanza typically reaches 6 to 8 feet tall and wide where it does not freeze back. When cut to the ground, it usually reaches 4 to 6 feet tall and wide by fall. Seedpods are sometimes produced, and seedlings may appear near your plants. Cuttings taken during the growing season are the way commercial growers produce more plants.

Close-up of **Hibiscus moscheutos** × *'Lady Baltimore,' which can be 10 to 12 inches in diameter. WCW*

› **Tecoma stans** *is often called "yellow bells" or "esperanza." This particular plant is blooming on a neglected site in Central Texas. WCW*

Close-up of a cluster of the bell-shaped flowers of Tecoma stans. WCW

Orange-flowered types tend to get larger and tolerate more cold than the commonly available yellow form. Like hardy hibiscus, yellow bells combine nicely with large-growing crinums and spider lilies. Rain lilies such as *Habranthus robustus* can be massed in front, but keep in mind the ultimate large size of the yellow bells.

Lantana is another good heat-tolerant plant, with the trailing purple form (*Lantana sellowiana*) making a beautiful combination with bulbs. All these plants need to be cut back in late fall or winter to make room for the *Narcissus* and *Leucojum* that fill in nicely during late winter and early spring. By the time the foliage starts to dry up on the bulbs, the hardy hibiscus, yellow bells, and lantana are beginning to start their activity.

Bulbs in the Evolution of a Garden

BY WILLIAM C. WELCH

# A Louisiana Country Garden: Mangham

Long before Chris was known as the Bulb Hunter and first began talking about the Southern Bulb Company, I suggested that he look at our property in Mangham, Louisiana, as a possibility for farming. I proposed that he stay there and meet some of our friends. Of course, Jessie Lee Harris was the first there to greet us, and he and Chris formed an instant bond. Jessie Lee had been collecting bulbs with me for many years; although he didn't know all their names, he knew all the important things about when they bloomed and how to grow them. Little did I know that Chris would be staying here often in the various seasons and that he and Jessie Lee would become well acquainted, dig bulbs together, and visit often on Chris's frequent trips.

Walking the long wooden hall and pacing across the cherry-wood floor of the kitchen with his cell phone held to his ear, Chris would often inquire about the house and the history of the gardens. There he was in Louisiana, and there I was at my home in College Station identifying plants and architecture from verbal descriptions. I do believe that Chris fell just as much in love with that home as Diane and I had.

Our first look at the property was in the summer of 1989. My late wife, Diane (1939–2007), and I were asked by her parents to look at it because they wanted a home in Mangham. The old family home on the nearby farm was too far gone to restore, and they needed a place to stay while conducting farm and ranching

◄ *An arch into the rear garden echoes the gables of the house.* Gladiolus byzantinus *bloom simultaneously with amaryllis in midspring. WCW*

The house and front garden just after purchase in 1989. WCW

business and visiting friends in northern Louisiana. We liked the house immediately. Built about 1900 as a merchant's cottage in Richland Parish in northeastern Louisiana, it was in reasonably good condition, and few changes had been made since the 1930s. The big living room, spacious hall, wonderful old screen door, tall ceilings, and generous front porch had great potential, although the bathroom and kitchen were badly out of date.

Most of all I was captivated by the property. It was right in the middle of town across the street from the Methodist church and Richland State Bank. A small but well-stocked neighborhood grocery store was right next door, and the owners even offered a charge account. Best of all I loved the remains of the garden. Laura Tillman and her husband ("Monkey") operated a store just down the street and had lived in the house for many years. She was a serious gardener. Old camellias thrived under towering pecans, and bulbs for every season were profuse and seemingly endless. Two huge specimens of Japanese magnolia (*Magnolia soulangeana*), a mammoth banana shrub (*Michelia figo*), and a ten-foot-tall sweet olive (*Osmanthus fragrans*) combined with drifts of 'Hyperion' (single yellow), 'Kwanso' (double orange), and *Hemerocallis fulva* (single orange) daylilies. The daylilies would provide a welcome, long season of color following the spring bulbs and flowering shrubs.

The front porch was a great place to view kids walking to school and a trail of farmers hauling equipment, cattle, and crops to town. A hedge of 'Formosa' azaleas rose in front of the porch so that if seated on the old metal-strap spring chairs, you could see out easier than others could see in. It was a fine spot to relax and enjoy the sights and scents of the community. When the banana shrub or sweet olive (also known as "tea olive") is blooming, the fragrance is wonderful. From the beginning, Jessie Lee was integral to the evolution of the house and garden. An enthusiastic and capable gardener, he understands plants and willingly shares them as well. Everyone in the community knows him and his wife, Mildred. A few old-fashioned hydrangeas grow in the garden, but Jessie Lee moved more from the farm. They were the

The Mangham house was built about 1900 and was typical of a small-town merchant's cottage. A mass of prairie phlox (Phlox pilosa) originated with a small plant originally from Elizabeth Lawrence's garden in Charlotte, North Carolina. WCW

old "mophead" types (*Hydrangea macrophylla*) that may be pink or blue depending on the acidity of the soil. Within two years of their "move to town" they were thriving and blooming in profusion. Later, David Creech from Mast Arboretum at Stephen F. Austin (SFA) State University in Nacogdoches shared oakleaf hydrangeas and others from his extensive collection. Greg Grant, also on the SFA faculty, brought summer phlox, verbenas, and dozens of other perennials, bulbs, and shrubs. Many friends have shared plants and cuttings to help build this garden. My bulbs had a special home. We planted numerous shared bulbs in our garden, but Chris was often interested in another plot of land that belonged to us and was located just down the road. Chris quickly became familiar with bulb plantings that Jessie Lee and I had established in rows near Diane's old family cemetery that dates to before the Civil War. Rows of 'Grand Primo' narcissus, snowflakes (*Leucojum aestivum*), and 'Grandjax' rain lilies attested to my earlier attempts at bulb farming. On one early summer visit, Chris took the time to wade through lush acres of knee-high corn

Pink and blue mophead hydrangeas (Hydrangea macrophylla) *thrive in the Mangham garden. WCW*

to discover rows of the pink rain lily (*Zephyranthes* × 'Grandjax') in full bloom. After planting those bulbs nearly two dozen years ago, we had not touched them, yet they had thrived all on their own.

Many of the other plants I chose for the cemetery and our gardens had similar abilities to thrive year after year, a term that would later be called "sustainability." I speculated that flowering dogwood trees would do well also, so my father-in-law, Norvel Thames, purchased some from the late Emory Smith, who became a famous native plant authority in Baton Rouge. Hilltop Arboretum is now part of the Louisiana State University (LSU) system and includes the former home of Emory Smith and his wife. The dogwood not only thrived but reseeded prolifically. I took a dozen one- and two-gallon container plants of these dogwoods to Chris last spring, who planted them at his bulb farm near Mineola, Texas.

The sharing of plants and gardening information in Mangham is alive and well. In the early 1990s my friend and well-known authority on old garden roses, Ruth Knopf, from Sullivan's Island, South Carolina, gave me a start of prairie phlox (*Phlox pilosa*) that she herself was given by the legendary writer and gardener Elizabeth Lawrence of Charlotte, North Carolina. I first planted them at Cricket Court in Washington County, Texas. When my father-in-law saw them in bloom, he asked for a few plants for the Mangham garden, where they have been a huge success. Jessie Lee bordered the ever-expanding island beds out front with Lawrence's phlox, and they created a great deal of attention, so much so that within two years they were being enjoyed by most of the other gardeners on Main Street. Chris met with Ruth Knopf in the heat of summer just outside Charleston, South Carolina, at Boone Hall Plantation. There they admired in particular the masses of white crinums (*Crinum powellii* 'Album') in bloom. Ruth has worked at the plantation for years and is now a consultant for its gardens. The earliest references to the land on which the plantation sits date to the late 1600s. She gave me an old tea rose she calls "Peach Tea," which thrives under dogwood trees in the front yard.

In order to best explain Chris's experiences with the Mangham

Drifts (elongated mass) of purply-pink prairie phlox (Phlox pilosa) provide at least two months of color each spring. WCW

gardens, I must mention more about the basic architecture, landscape design, and plants that build the foundation of the garden. One of the most important of these is the collection of deciduous shrubs, such as forsythia (*Forsythia × intermedia*), flowering almond (*Prunus glandulosa*), mock orange (*Philadelphus coronarius*), and three kinds of spirea. Baby's breath (*Spiraea thunbergia*), popcorn (*S. prunifolium*), and bridal wreath (*S. cantoniensis*) begin flowering in late winter and continue through late spring. Flowering almond was there originally, but I was able to add the white form from Mrs. Childress's garden. She was the local home economics teacher and a great gardener. I remember her for generously sharing pink Roman hyacinths and other garden delights but also for saying, when asked how she was able to get gardening assistance, "It's simple; you pay them!" She also had a peony collection that fascinated me.

Bobby McDonald, a retired hospital administrator who moved to Mangham from California, had built a duplex-style home surrounding a courtyard to share with his mother. He brought with him a passion for gardening, especially for azaleas, camellias, and narcissus. Bobby generously agreed to bud some of his best camellias onto seedlings that had volunteered in our garden.

My first specimen of Chinese fringe tree (*Chionanthus retusus*) was from the late Lynn Lowrey in about 1978. While visiting with Lynn at his nursery in Houston, I helped unload plants from Tom Dodd's Nursery in Mobile, Alabama. Among the delights was a group of about a dozen *C. retusus* blooming bravely in one-gallon containers. I was surprised when my plant did well in College Station, and the more I observed it, the more enthusiastic I became. The late Aubrey King, a third-generation nurseryman at King's Nursery in Tenaha, Texas, liked a propagation challenge and was able to root cuttings from my tree and offer a few for sale. Mitch Goyne, a graduate student at Texas A&M University at the time and a native of nearby Rayville, Louisiana, was also able to root a few cuttings. I was fortunate to get two five-gallon container plants from Aubrey for the Mangham garden. Although a close kin of the native fringe tree or grancy graybeard (*C. virginiana*), the Chinese version is even more useful in my opinion. Laura Tillman had planted what is now a large specimen of the native species on the west side of the front yard, so we can compare the two.

Any one of these many shrubs could be blooming when Chris arrived. After our initial visit I suggested that Chris meet and talk with some of our longtime friends in Newellton, Louisiana, about fifty miles east in the Mississippi River delta. The bulbs in that area are particularly plentiful because of the deep alluvial soil that has continued to support crops of cotton, corn, and soybeans for many generations.

These were the magic words for Chris, and he asked if he could stay for some extended time in the house as he traveled many of the roads in the area looking for bulbs. We began to talk daily about his hunts; I would ask about how the gardens looked and about the house. Periodically I would have Chris check on small projects done on the house in my absence, but the home in which Chris now stayed was not what it had once been.

Renovating the House

In 1999, we decided to do some restoration work on the house. I wanted to give some structure and organization to the garden and added walks, landing areas at the front and rear entrances, two wooden arches, and cypress picket fencing to define the garden.

We were encouraged by an LSU forester to harvest some black cherry (*Prunus serotina*) and bald cypress (*Taxodium distichum*) trees from other property we owned because they needed thinning. The cypress was harvested from the Buie property, which has been in my wife's family since before the Civil War. The cherry was from a twenty-acre woodlot on Little Creek, close to the old Thames homesite (Thames was the maiden name of Diane, my first wife). Leo Miller, from Round Top, Texas, brought his portable sawmill and stayed several months cutting and

Coordinating color displays with bulbs, shrubs, and other plants is an art that is learned over time. WCW

milling the lumber. It dried in an open barn on the farm and was used as the project unfolded. Much of the cypress became picket fencing and decking on the porches; and the cherry, flooring and a mantel in the small addition to the back of the house.

Chris was interested in lumber harvested from the original woodlot of our property and used in the restoration of the house. Several years later he asked to see some of the unused cut wood. When we arrived at Mangham that afternoon, he noted the beds of bulbs Jessie Lee had planted under old pecan trees in the yard.

A very simple but powerful combination of single and double orange daylilies, red spider lilies (*Lycoris radiata*), surprise lilies (*L. squamigera*), and *Narcissus tazetta* 'Grand Primo' was planted in a circle around the base. Whenever one plant was dormant, another plant was growing, and there were very few seasons in this bed without something in bloom. Chris was inspired, and his excitement spread to me. We would later spend hours putting together other creative perennial bulb combinations.

It seemed that every season offered naturalized bulbs that I now consider "buried treasures." Early and midspring the jonquils, narcissus, Roman hyacinths, and snowflakes emerge in bursts of color and fragrance. These are followed by the Johnson's amaryllis and a large number of amaryllis 'Apple Blossom' (*Hippeastrum* 'Apple Blossom'). My coworker Cynthia Mueller in College Station is a breeder and collector of amaryllis and has shared a number of others. The late Cleo Barnwell of Shreveport shared her collection of *Lycoris* and Spanish bluebells (*Hyacinthoides hispanica*). I was delighted to find a few white Roman hyacinths scattered about the garden and surprised that they sometimes bloom as early as Christmas. Their fragrance is unsurpassed, and they appear when few other plants are blooming. Flora Ann Bynum, from Old Salem, North Carolina, was the best grower and authority of Roman hyacinths that I know. Flora Ann and Greg Grant pronounced judgment on the authenticity of my somewhat limited but varied collection. We lost our dear friend Flora Ann in the winter of 2007, leaving the hyacinth identification burden to Greg, who cheerfully pronounces his opinion when asked. A nice gift of hyacinths came from Margaret Ann Thames's garden down the street and some from Virginia Sue Barr in Oak Ridge, Louisiana (about twenty miles north of Mangham), and I brought a few blue and white ones from the Round Top area. Mrs. Childress had shared several of her rare pink hyacinths with this warning: "They must be divided and reset every three or four years." Florence and Bill Griffin of Atlanta had taken us on a bulb hunt in rural Georgia, and a gracious gardener had shared some of her pink hyacinths with us. Among the joys of gardening in Mangham are the narcissus. They begin with some early paperwhites (*N. t. papyraceous*) followed by masses of *N. t.* 'Grand Primo,' sweeties (*N. jonquilla*), *N. t. italicus*, Lent lilies (*N. pseudonarcissus*), "Butter and Eggs" (*N.* 'Orange Phoenix'), and 'Campernelle' jonquils (*N.* × *odorus*). By late February large masses of 'Grand Primo' dominate the parade. Jessie Lee has continued to add narcissus to the island beds in the front yard in spite of my pleas for moderation. Snowflakes have a relatively long bloom season starting in early February. Hardy amaryllis (*Hippeastrum* × *johnsonii*; sometimes referred to as *H.* 'Johnsonii') also bloom for about a month near the end of the spring season. The snow-

Narcissus 'Sir Winston Churchill' is the latest-flowering narcissus for much of the South. It is fine in form and has a crispness of appearance that adds to any garden. WCW

Chris proudly displays a large stem of Crinum *'Mrs. James Hendry' that was blooming profusely in the August heat WCW*

flakes came with the garden but have increased rapidly because Jessie Lee occasionally divides them. As spring progresses, the Byzantine gladiolus (*Gladiolus byzantinus*) and later the parrot glads (*G. dalenii*) continue the spring display. Rows and drifts of *N.* 'Golden Dawn' conclude the narcissus season. Competing for attention while all this is going on are bearded and Louisiana irises. 'Caesar's Brother,' a deep blue-purple Siberian iris, has also thrived for us and was a gift from Mattie Rosprim, a local gardener in College Station. Under one of the huge Japanese magnolias on the west property line is a clump of tiger lilies, a remnant of Laura Tillman's garden. Although threatened by the increasing shade, they continue to flower in early summer and produce the amazing little bulbils.

I had never seen spider lilies thrive as they do in northeastern Louisiana. After the first good early fall rain, many vacant lots and lawns are thick with them. We have combined spider lilies with single and double orange daylilies, 'Grand Primo,' and *N. t. italicus* in circular plantings under the old pecan trees in

A vase filled with *Crinum powellii.* 'Album' is one of the best and most prolific crinums available. They not only look good in the vase but are fragrant and very easily grown. WCW

the backyard. Chris and Jessie Lee have dug large numbers of them in the area. Our old pecan trees drop their leaves each fall, providing plenty of sun for the foliage and bulbs to prosper. In July or August, *Lycoris squamigera* appear as if by magic after a summer rain. Known as "surprise lilies" or "naked ladies," their two-foot-tall stems of bright pink flowers are truly elegant in the summer heat. Virginia Sue Barr has the most impressive planting of them I have seen, surrounded by the circular drive at the end of an allée of pecan trees.In 1999, I moved most of my crinum collection to Mangham. These were mostly from Scott Ogden, and I had enjoyed them for many years in College Station until we built a new garden home and no longer had room for all of them. They are enjoying the good soil and additional rainfall, although the new site is a bit too shady for them to bloom as much as I would like. I love crinums for their ability to survive heat, relatively poor soils, and pests. 'Mrs. James Hendry' is pale pink and repeats from April through November. Its fragrance is outstanding as well. I planted about a dozen blooming-size bulbs in the center of a thirty-foot-square bed in the backyard. Two years ago I gave most of the crinums to Chris so that he could share them with his customers through the Southern Bulb Company. Luther Burbank's 'White Queen' and *C. powellii* 'Album' are my favorite whites. 'Bradley' (dark pink), 'Carroll Abbott,' 'Mardi Gras,' and 'Ellen Bosanquet' are also longtime favorites. Years ago I had given a couple of blooming-size 'Carroll Abbott' bulbs to Mary Anne Pickens for her garden near Columbus, Texas. My two clumps at Mangham were blooming and multiplying nicely when Chris spied them and longingly commented on how much he would like to get them into production. I relented and also threw in most of my stock of 'Mrs. James Hendry.' I realized last year that I was really missing 'Carroll Abbott' with its continuous flowering. Mary Anne recently sent me two large bulbs, and I'm not telling Chris where I planted them!

Mayhaws

The first trees to bloom in the Mangham backyard each spring are the mayhaws (*Crataegus opaca*). I love mayhaw jelly! We all looked forward to late spring when roadside vendors would sell the jelly. In about 1991, Diane's dad decided to visit the late nurseryman Sherwood Akin in Sibley, Louisiana, who was famous for his knowledge and production of selected mayhaw varieties.

Mayhaws are worth planting just as ornamentals. The handsome trunks, early white spring flowers, and crab apple–like fruit adorn a small tree about the size of a redbud or dogwood. Native to acidic soil areas of the Southeast, they prefer moist areas and sometimes occur in swamps, where the fruit can be scooped from the water with nets. They do just about as well in well-drained soil, and bulbs planted around their bases work beautifully. 'Campernelle' and 'Grand Primo,' along with snowflakes, are the bulbs of choice for spring with a healthy sprinkling of *Lycoris* for the fall.

We purchased three different grafted selections of mayhaws. One was planted in town and two at the Buie cemetery, about a mile from my home in Mangham, Louisiana. All three continue to prosper, and in some years Jessie Lee is able to gather large quantities of fruit. Mayhaws store easily in the freezer, and I am able to make a couple of batches of jelly from a good crop. Unfortunately, they are often victimized by rust.

Fig Season

I have been glad to share my home, plants, and heritage to help Chris with his numerous Bulb Hunter projects. At one time, he and a film crew used the gardens in the summer to film a potential pilot for a TV program. The only bulbs in full bloom at the time were the Philippine lily (*Lilium formosanum*) and the surprise lilies. I think what everyone enjoyed most of all was not the bulbs! Along the back fencerow of the property is a line of plantings of some of my favorite figs.

Figs (*Ficus* spp.) are a joy each summer in Mangham. The local favorite is 'Celeste,' sometimes known as the "Texas sugar fig." Occasionally, the plants are damaged by cold, which causes them to send up numerous trunks. The single-stemmed specimens are what I remember from my childhood in Houston. Laura Tillman must have loved figs because several "clumps" of them grew in the garden. I removed some but still have an area about twenty by thirty feet that is a mass of figs each summer. I do enjoy the variety of having several different figs. I like the dark purple 'L.S.U. Purple' and the huge, yellowish fruit of the later-ripening "South Carolina lemon fig," which is among my favorite plant gifts from Frances Parker in Beaufort, South Carolina. Aubrey King and Heidi Sheesley have propagated it so that others can enjoy them each summer. 'Celeste' and 'Texas Everbearing' are still the favorites among most Texas and Louisiana fig aficionados.

Phlox

In addition to the Elizabeth Lawrence *Phlox pilosa* and the local "Gaudy Pink" selection of summer phlox (*P. paniculata*), a new favorite is Greg's introduction 'John Fanick' with its lavender-pink centers surrounded by white. It is the most vigorous and productive summer phlox I have ever grown. I love the color combination of "Gaudy Pink" planted with 'John Fanick.' They are also great cut flowers and have a delightful fragrance.

Roses

Having found and named a rose at the original farm site for Diane's grandmother ("Maggie") and having finally identified the swamp rose (*Rosa palustris scandens*) from that location, I added some others to the Mangham garden. An all-time favorite is the "Natchitoches Noisette" that I found at the American Cemetery in Natchitoches (nak-uh-tish), Louisiana, in the early 1980s. Another favorite is a found rose from Ruth Knopf she calls "Peach Tea." I decided to use it as a specimen in the large bed to the right of the front steps. Most of the site is too shady for roses, but another favorite, 'Marie Pavie,' does well to the right of the rear entrance.

Odds and Ends

Jessie Lee added old-fashioned balsam (*Impatiens balsamina*, also called "lady's slipper," not to be confused with the orchid) in pink and purple from his mother's garden. They reseed happily. Each fall I look forward to the Japanese anemones (*Anemone hybrida*), a gift from Jason and Shelley Powell of Petals from the Past Nursery in Jemison, Alabama. I had first seen them in the late Margaret Kane's garden in San Antonio but have really enjoyed the pink and white forms from Jason. They are a welcome addition to the fall garden.

My introduction to sweet shrub (*Calycanthus floridus*) came from the late Catherine Sims in Homewood, Alabama. Catherine declared hers had the largest and most fragrant blooms of any she had seen. Its burgundy flowers adorn the back fenceline each spring. When Catherine, known as the "Plant Lady of Home-wood," passed away, she donated her property to the city. Those properties opened in 2012 as a city botanical garden known as "Sims Ecoscape."

Soon after the renovation of the house was finished in 2000, we drove into Mangham late in the evening. As the headlights shone into the backyard, I noticed something different. Realizing that maintaining additional planting areas was a growing challenge, I had asked Jessie Lee to concentrate on the existing beds and try to limit "new ones." As we got closer to the area, we could see old white-painted brick in a sawtooth edge of a garden, and Diane said, "This is wonderful! A knot garden!" and I added, "Yes, a "not" garden.

Camellias

One evening, about an hour before sunset, I received a call from Chris announcing that he just happened to be passing through the northern Louisiana area and asked if he could stay at Mangham. I said of course and encouraged him to call Jessie Lee.

It was a great time for Chris to stop and a time that I rarely have the ability to visit myself. The cemetery white irises were blooming among the purple verbenas under the large Japanese magnolias with the 'Golden Dawn' narcissus beneath them. The white flowering dogwood were unfurling their blooms in all of their glory. I told Chris that Diane's dad really liked the pink dogwood, but they didn't reseed and thrive like the white ones he had purchased from Emory Smith in Baton Rouge. I asked Jessie Lee to send Chris home with a dozen of the two-foot-tall gallon-and two-gallon-sized seedlings he had been growing under the pecan trees. On the tail end of their bloom, Chris saw some very old camellias in bloom.

I was delighted to find large plants of many of the old varieties of camellias in the Mangham garden. Their bloom season is long and begins with the hedge of sasanquas on the west side of the backyard, where they bloom in November and December. A six-foot, irregular border of *N. jonquilla* starts sending up its reedlike

Old specimens of camellias thrive on the property. This Camellia 'Rose Dawn' is providing a beautiful display during mid-March. WCW

foliage by November. Favorite camellias include 'Purple Dawn,' 'Pink Perfection,' 'Debutante,' 'Rose Dawn,' 'Chandleri Elegans,' and more recently as a prized gift from Lynn Stiles, the beautiful white 'Alba Plena.' After having struggled with growing camellias in Houston and College Station, I find it delightful to have them in Mangham, thriving and reseeding.

Southern Living

One summer I invited *Southern Living* garden editor, Gene Bussell, photographer Ralph Anderson, and Chris to visit my friend Virginia Sue Barr, whom I introduced earlier. Our goal was to see the "naked ladies" (*Lycoris squamigera*) in full bloom. We weren't disappointed! If you recall, Chris had called on Virginia Sue earlier in his career when her blue Roman hyacinths were in full bloom. We also have lots of "naked ladies" in Mangham, as well as the more common and later-blooming red spider lilies.

The entire community became excited when *Southern Living* magazine decided to photograph the garden and write an article. The local quick-stop grocery store and filling station had "Welcome Southern Living" on their marquis, and everyone in the community slowed down and watched what was going on.

Jessie Lee had been working toward the April date for weeks, and several friends offered to come help. Among them was Lynn Stiles. Lynn was from Newellton but went to school in Texas and worked for the Texas Agricultural Extension Service for a time while her husband, David, completed his degree at Texas A&M. When she offered to assist, I was pleased but didn't fully comprehend what she could do.

We were starting a day's work when she arrived, and I asked if she would help plant some gallon-sized old-fashioned petunias I had brought from King's Nursery. Jessie Lee offered shovels and hoes. Lynn responded, "The handles on the wooden tools are too fragile. I have my own metal ones!" Despite her petite size, Lynn set the pace for physical work. She pruned, planted, weeded, mulched, and did any other jobs we had. When Gene and Ralph arrived, they, too, pitched in. That evening we went out to dinner at a local restaurant known as "Sister Baby's," located in a nearby cotton field. The menu included several seafood items and typical Louisiana fare such as choice steaks. The waitress was fun and full of life. Ralph was the last to order, and she told him, "Our steaks are really good, and nobody has ordered one. Can I get one for you . . . big boy?" Ralph was a bit overcome and agreed. The steak that he was served was really large. We took home enough for all to have steak sandwiches the next day.

Ralph started to photograph just before daylight the next morning, and traffic slowed to a crawl, as everyone wanted to see what was happening. Ralph used deflection devices for the light and various other equipment not commonly seen in Mangham. He is a very talented guy, and I think he and Gene enjoyed the experience of being in Mangham and absorbing the local culture.

Traditions Continue

Although I spend less time in Mangham than I would like, the garden continues to draw me there. Friends like Neil Odenwald stop by frequently to visit with Jessie Lee and gather bouquets or bring something new to plant. Neil shared his wonderful Taiwan cherry trees (*Prunus campanulata*) from Baton Rouge as well as a nice specimen of snowball (*Viburnum macrocephalum* 'Sterile'). On a recent visit to the Mangham garden the bulbs were in full bloom, and Neil commented, "I like the winter scene. Island beds have a commanding presence in the front garden." Newellton friends Helen Hester Kifer, Lynn Stiles, Jan Burnside, and others pull in for brief visits on trips to and from Monroe. Graduating seniors from Mangham High School have their pictures made in the garden, and the Mangham United Methodist Church has its annual Easter egg hunt there as well. Jessie Lee asks frequently when Greg is coming back to play the piano and Chris to dig bulbs. My son, Will; his wife, Mandy; and beautiful daughters, Alyse and Ella, love their visits to Mangham, where Will remembers good times there with his grandparents. I continue to be inspired by the seasonal change and color in the garden and all those from the past and present who are represented there.

On a more recent visit Chris and I were traveling together for a talk he was to give in Shreveport. We overnighted in Mangham afterward and enjoyed the abundance of 'Golden Dawn' narcissus in bloom. They signify the close of narcissus season, but 'Sir Winston Churchill' presents the final display.

I offered the crinum collection I had moved from College Station to Chris, and he eagerly accepted. Chris, Jessie Lee, and I dug them on this trip, and Chris planted them at his bulb farm. My granddaughter, Kathlyn Alyse Welch, loves to smell crinums, especially my favorite, 'Mrs. James Hendry.' The wonderful thing about crinums is that we know that they will all be thriving long after we are gone.

We lost Diane in a heroic battle with melanoma in April 2007. In August 2009 Lucille Presley and I were married in a beautiful ceremony at our church in College Station. Lucille and I share many things, including our successful marriages and lifetime love of gardening. Last summer I began gifting her with plants from the Mangham garden, including Emory Smith's dogwoods and Elizabeth Lawrence's prairie phlox, along with Greg Grant's 'John Fanick' phlox. In March, Jessie Lee carefully dug and potted two root sprouts from the old specimen of 'Pink Perfection' camellia. Lucille's garden, "Twin Oaks," at Independence, Texas, has a magnificent view and includes three live oaks that are three

> Phlox paniculata *'John Fanick'* has become a star performer in the summer garden. This photo was taken at Mangham, where it gracefully adorns the walk at the side entrance to the house. Greg Grant introduced the plant. Its vigor, fragrance, and long summer bloom season may become a winner all across the South. WCW

to four hundred years old. Chris and Greg both gave us narcissus bulbs in celebration of our wedding. Chris visits Twin Oaks regularly, and we have planted lots of his narcissus and rain lilies recently. Lucille and I are enlarging a vegetable garden at Twin Oaks and planning other ventures there as well. I am quickly adding crinums (including the two new 'Carroll Abbott' from Mary Anne Pickens). We hope that gardening will continue to be a sharing experience with old and new friends that will keep us young. We are also hopeful that Kathlyn Alyse will share with her younger sister, Ella Diane, the natural gardening interest she has. Sharing our gardening heritage has become a passion we hope to pass along as long as possible.

A Texas City Garden: Pebble Creek

I n 1989, my late wife, Diane, and I wanted a smaller and lower-maintenance place in College Station. The garden home section of Pebble Creek seemed ideal. College Station is in the heart of a triangle between Austin, Houston, and Dallas–Fort Worth—major cities in the state of Texas. One can imagine then that Chris would become a frequent visitor. To

Iron obelisks support "Peggy Martin" roses in a bed filled with bulbs and perennials. WCW

Rubeckia ×
'Herbstonne'
makes four-foot-
tall clumps of
yellow, daisylike
flowers for about
two months
each summer.
It is a great
companion
perennial for
crinums and
Hymenocallis.
WCW

better understand the garden he would often stroll through, and eventually take bulbs from, I need to share a few details.

The front yards had quite a few restrictions to make them all fairly compatible, but the fifteen-foot side space and twenty-five- by eighty-foot strip along the golf course offered sufficient space for creativity. The front spaces are all maintained by the neighborhood association, which also changes some color plantings at each home twice a year. I decided to employ a friend and local landscape architect, Bob Ruth, to assist in planning the hardscape, especially the drainage, vehicular circulation, and construction details.

Now, let us move on to the bulbs that would eventually end up in Chris's hands. For foliage and flowers I added an eight-foot-long drift of *Crinum jagus* 'rattrayi' as well as a smaller planting of *C. jagus* 'scillifolia,' both commonly called 'scillifolia' and 'rattrayi.' 'Scillifolia' is very similar to 'rattrayi' except for having narrower leaves. These two crinums have the best-looking foliage of any in the genus. They could be planted just for their broad, dark green leaves, but each year in late June or early July they have huge, bulbous, vanilla-scented flowers that dominate the garden for a couple of weeks. These have cheerfully survived Chris's division process over the past several years. A large clump of spider lilies (*Hymenocallis* 'Tropical Giant') adds dozens of white flowers to the summer garden and contrasts nicely against a hedge of dwarf Burford holly (*Ilex cornuta* 'Burfordii Nana').

Perhaps one of the large values this garden brought to Chris's horticulture growth was that it gave him the ability to see the importance of using other perennials mixed with bulbs for a more mature garden. A big specimen of *Hibiscus moscheutos* × 'Lady Baltimore' blooms with eight- to ten-inch disks of pink from a mass of dwarf nandina, creating a nice display summer through fall. Good friends and neighbors Martha and Gerald Still, along with Mike and Lou Anne Workman, asked if I could root them a cutting, and Cynthia Mueller volunteered to do so. I hadn't planned on it becoming a neighborhood competition, but both these normally friendly couples declare that they have won the 'Lady Baltimore' contest, and I must admit that theirs both look better than mine. I trimmed around mine last week, gave them a good dose of cottonseed meal, and hope that I can regain my reputation. My wife, Lucille, has requested a couple of plants of 'Lady Baltimore' and 'Moy Grande' (even bigger but solid dark rose). I took cuttings from an old plant set out many years ago at Cricket Court, near Winedale, Texas, and set them in pots under some azaleas at Twin Oaks. They appear to be getting enough water from the sprinkler system and should be ready to set out in a couple of weeks. The Stills and the Workmans soon will have new competition! Chris has also taken cuttings and begun growing them on the farm.

We had wanted a gazebo-like structure in the center of the side garden, which was a challenge since solid buildings were not allowed in that space. Bob and I came up with an ornamental iron structure and capped it with an antique finial. It seemed appropriate that it should be covered with Lady Banks roses. The neighbor's house wall was softened with freestanding trellis work, offering opportunity for more climbing roses and other vines. The area beneath the gazebo has become rather shady, but it is interesting that my favorite rain lily, *Habranthus robustus*, reseeds and blooms furiously there several times each summer. Cynthia Mueller provided the start, and now they are quite prolific.

Chris harvested a significant amount of these particular rain lilies from this location, and we often discuss them when he visits here. We first note how they hardly appear to have diminished at all, despite the large number taken out. Second, we note that over the last eight years, they have bloomed much better in some years than others, a phenomenon that we cannot quite explain. Last, it is interesting that Chris grows them in complete sun where they do well for him, but these bloom better here in Pebble Creek under the gazebo, which could be considered quite shady, with about half a day of filtered sun. They are the perfect selection for that spot in my garden and offer a nice splash of color in this transition area to the front gardens.

The major plant specimen in the front yard is a large Chinese fringe tree that we moved from our previous garden in College Station. The original was grown by Tom Dodd Nursery in Semmes, Alabama, and was given to me by Lynn Lowrey. A planting between the driveway and the dining room with its tall windows included a low hedge of dwarf yaupon, several 'Marie Pavie' roses, and 'Grand Primo' narcissus, a gift from Greg Grant. We held our breath the first year and were delighted when the Chinese fringe tree began to flourish. The other specimens, a ten-foot-tall 'Yvonne' crapemyrtle and a 'Changsha' tangerine were also moved in, which added some "instant maturity" to the front garden. A nice specimen of *Clethra pringlei* (commonly called

A large specimen of Chinese fringe tree (Chionanthus retusus) creates attention each spring when it blooms. It was planted at our previous College Station home from a one-gallon container in 1979 and moved to Pebble Creek in 1999. WCW

"Mexican sweetspire clethra") was a gift from Brad Jennings, who was a student in landscape architecture at the time. Brad built a beautiful sawtooth brick edge along the golf course walk. Boone Holladay, another graduate student in horticulture, built a beautiful brick platform to display a large glazed pot that held an 'Alba Plena' camellia.

The theme of heirloom gardening is strong in this garden, and many heirlooms are passalongs. This means that every plant, every fruit, every accessory has a history. Chris soon discovered this, and we would often find ourselves strolling through the garden recounting the stories of each plant or garden accessory. Chris asks questions about this or that, and I know I must have told him the story at least four or five times already. Either Chris has a terrible memory, or he simply likes hearing the story again.

The entrance court was enhanced by a recirculating fountain featuring an antique cast-iron lavabo (a water-retaining device traditionally used to wash one's hands), also moved from the earlier garden. Jason and Shelley Powell, owners of Petals from the Past Nursery in Jemison, Alabama, contributed a 'Brown Select'

Clethra pringlei *is a rare, evergreen shrub from Mexico that has beautiful racemes of flowers each year in July. WCW*

Containers of 'Laura Bush' *petunias and various succulents add color and interest without requiring high maintenance. WCW*

satsuma and two 'Meiwa' kumquats, which are in the protected environment of the side court. The satsuma produces a nice crop every year, but in alternate years it may have up to two hundred fruits. Our granddaughter Kathlyn Alyse loves them and likes to pick a few extras when they are ripe to pack in her school lunch. Jerry Parsons provided rooted plants of Greg's 'Marie Daly' roses as well as the dark pink coral vines (*Antigonon leptopus*).

The four original teakwood and iron obelisks are covered with "Peggy Martin" roses. After about eight years the teak rotted and I had them rebuilt with all iron. The area between them has 'Belinda's Dream' roses, 'Mrs. B. R. Cant,' 'Souvenir de la Malmaison,' and a mixture of annuals and perennials, including Greg's wonderful *Salvia farinacea* 'Henry Duelberg,' *Phlox paniculata* 'John Fanick,' and *P. pilosa* (originally from Elizabeth Lawrence's

The front court has a large specimen of 'Prosperity' rose (Rosa 'Prosperity'), an antique fountain, seasonal bulbs, and perennials. WCW

Spineless forms of prickly pear are very easy to grow and offer spring flowers and summer fruit. Here a coral vine (Antigonon leptopus) gracefully drapes itself after tumbling over a nearby fence. Coral vine is an easy-to-grow perennial that flowers from late summer until fall. WCW

garden by way of Ruth Knopf). My two favorite sweet peas are 'Painted Lady' and 'Cupani.' Both reseed faithfully on the fence and on top of the prostrate rosemarys. There is a very narrow strip along the golf course walk that Chris suggested I plant with oxblood lilies and rain lilies. There are also several roses trained on the fence, including "Peggy Martin," 'Clair Matin,' and 'Cl. Blossomtime.' They are heirloom types and provide many bouquets of fragrant cut flowers each spring.

Chris is a frequent visitor and always "checks out" the masses of *Crinum* 'Mrs. James Hendry' that anchor the long planting bed along the golf course. The first time he requested divisions, there were quite a few in the clump on the north end. We took some of

On the porch level a sixty- by three-foot-wide bed is filled with heirloom narcissus, Gladiolus byzantinus, *'Marie Daly' roses, garden mums, and seasonal annuals. It is also bordered by white rain lilies* (Zephyranthes candida) *that Chris has "divided and reset" on occasion. WCW*

An iron arch provides support for 'Reve d'Or' roses against the neighbor's house wall. In the foreground two large plants of 'Belinda's Dream' roses, a mass of Phlox paniculata *'John Fanick,' and* Crinum *'Mrs. James Hendry' (not in bloom) provide color and fragrance to a much-used part of the garden. WCW*

the smaller ones and created a "drift" of seven at the south end of the bed. Chris times his requests for more 'Mrs. James Hendry' divisions carefully. The requests usually occur after the second round of drinking old fashioneds on the porch. Crinums continue to increase in popularity, and those that repeat-flower and have a really good fragrance are higher on my list of favorites. I have particularly enjoyed a growing clump of 'Rose Parade,' a gift from crinum specialist Steve Lowe. David Dement, a Master Gardener leader from Gonzales, gave me a large bulb of 'Super Ellen' last year. After taking two years to get established, it has rewarded us with multiple flower stems. They are a medium pink, and the flowers are unusually nice for cutting. As the name implies, 'Super Ellen' is especially vigorous and large growing (four to five feet)

A summertime favorite are the dozens of Philippine lilies that come back from bulbs and also reseed in the upper and lower parts of the rear garden. The prostrate rosemarys soften the long retaining wall where "too many" potted specimens add seasonal interest. Last spring Dan Lineberger, professor of horticulture and longtime friend, tissue-cultured 'Laura Bush,' my favorite petunia, and offered them for sale at the student plant sale event. Lucille and I were about to host a party for Chris and Rebecca's wedding announcement, and I purchased several dozen six-inch pots in full bloom. Their color is almost identical to that of the beautiful spikes of *Gladiolus byzantinus* that were just beginning to show color. 'Laura Bush' is very much like the pink, purple, lavender, and white old-fashioned petunias that reseed in my gardens each spring. As I write this, the petunias are declining but still colorful. Greg had given me a start of his *G. byzantinus* several years ago, and Chris traded several dozen large corms for the 'Mrs. James Hendry' crinums he had "liberated." Also in flower at that time is one of my favorite narcissus, 'Sir Winston Churchill,' which is the last one to bloom for me. *Narcissus tazetta* 'Grand Primo' and 'Golden Dawn' have been added from the Mangham garden. The narcissus adapted to our area are most welcome when they begin flowering in late winter and early spring.

A rainfall capture system is a big asset to success with container plantings in College Station. The municipal water is high in sodium and results in damaged plants during long spells of

Crinum 'Super Ellen' is prospering in the long border. Here, two stems have been cut for us to enjoy in the dining area of the house. WCW

Philippine lilies (Lilium formosanum) *appear on three- to five-foot stems in midsummer when little else is blooming. They are reliable perennials and also reseed in the garden. WCW*

A six-foot row of Gladiolus byzantinus *from Greg Grant makes a bold color mass in midspring. WCW*

limited rainfall. To conform with neighborhood restrictions, we could not use aboveground cisterns, so we connected the rain gutters underground with two tanks (six hundred gallons each) under the paved terrace. Water is then moved by a submersible pump to a convenient tap where water cans await filling and distribution to the plants. Rainwater not only helps to conserve our water resources but also provides a pure water source for camellias and a few ferns and orchids that are highly sensitive to sodium.

Chris and I often find time to relax on the back porch in rocking chairs, a gift from friends who traveled with us to Provence the summer of 1999 when we were building the house. They continue to remind me of the wonderful time we had as we explored the restaurants and gardens of that region. A favorite handmade hypertufa pot and plant is a gift from Joe and Beverly Tocquigny in Seguin (*Myrtus communis* 'Compacta,' known as a compact myrtle). Hypertufa is a collection of aggregates held together by a type of cement known as portland cement. A large potted specimen of *Camellia* 'Delores Edwards' performs well. Cynthia rooted two cuttings of Elizabeth Lawrence's 'White Empress' camellia, which is the most vigorous and satisfying camellia I have ever grown. Thanks to Lindie Wilson for being such a good steward of Elizabeth Lawrence's plants and gardens and for sharing them so generously with me over the years. Thanks also to Patti McGee, who has worked so tirelessly to preserve that garden through a cooperative agreement with Winghaven Gardens in Charlotte and the Garden Conservancy.

The rocking chairs are wonderful, but anybody who knows Chris knows that he can't sit still for long. He is also a master at managing others to help him work. So, as one might imagine, the next morning after a leisurely evening, I found myself up before dawn helping size and sort a small batch of Byzantine gladiolus that he had brought in his truck. As wages for my labor I asked for the reject bulbs that could not be sold, and I planted another row of Byzantine gladiolus along my bottom fence.

Container plants are fairly important in the Pebble Creek garden. I tend to plant more of them in the spring than I care to water in the summer! We especially enjoy succulents that have a nice form, texture, and flowers. *Sedum potosinum* doesn't

Sedum potosinum *blooms starry white flowers for a month or more each spring and has evergreen bluish-green foliage. It is easily started from cuttings and an heirloom in Central and South Texas gardens. WCW*

have a common name, but I have grown it since a child in Houston. Its grayish-blue foliage is compact and especially nice with the white, star-shaped flowers that bloom in May and June. It is effective when used to skirt the base of larger plants since it rarely exceeds three or four inches tall. *Sedum palmeri* was first received from plantsman Scott Ogden and is another favorite. It reaches about six to eight inches tall and blooms for a month or more each spring with clusters of yellow flowers. *Sedum* 'Ogon' is relatively new on the market and is popular for its chartreuse foliage. I have particularly enjoyed it in an antique *faux bois* wall planter purchased from Joshua's Nursery in Houston last year. This past winter I enjoyed several pots of *N. t. italicus* purchased from Chris's company. I'm always on the lookout for plants that will combine well with bulbs. The herb thyme seems to be compatible with some blue Roman hyacinths in not liking much water in the summertime. I also have several pots of *Graptopetalum paraguayense*, better known as "hens and chicks" or "ghost plant." All these succulents root easily by placing tip cuttings in pots of fresh soil. A striped form of *Agave americana* from Lynn

Sedum palmeri is an easily grown plant from Mexico that blooms for a month or more each spring and requires little watering in summer. New plants may be started from tip cuttings at any time of year. WCW

Succulents like "hens and chicks" (Graptopetalum paraguayense), Sedum potosinum, and S. palmeri provide color and texture with only occasional watering. They are easily started from cuttings at any season of the year. WCW

Lowrey flourishes in the green space just outside our yard. These easy-care plants become really important during our hottest months of the year when other annuals and perennials are "taking a rest."

It seems that summer is when I see Chris here the most. The cabin and farm, although home to him, lack proper cooling facilities. Just recently Chris began to grow some of the succulents at his cabin that also adorn the landscaped areas here. He has, of course, underplanted each pot of his succulents with bulbs, such as rain lilies (*Zephyranthes grandiflora*) and spider lilies (*Lycoris radiata*). The Pebble Creek house and garden continue to be home for Lucille and me and, like all my gardening experiences, a place to learn and enjoy plants, landscape compositions, and good friends.

Index